LAKE HOPATCONG

LAKE HOPATCONG

A HISTORY OF NEW JERSEY'S LARGEST LAKE

PETER ASTRAS

Published by The History Press
Charleston, SC
www.historypress.com

Front: Lake Hopatcong in Sparta, New Jersey. *Photographs in Carol M. Highsmith's America Project in the Carol M. Highsmith Archive, Library of Congress, Prints and Photographs Division.*

Back (main): Lake Hopatcong Yacht Club dock, circa 1910. *Library of Congress, Prints and Photographs Division.*
Back (inlay): Sandi Astras. *Personal collection.*

First published 2024
Updated printing 2024

Manufactured in the United States

ISBN 9781467154710

Library of Congress Control Number: 2023947103

To Sandi Astras for her invaluable assistance with this book's art and photography. Her steadfast support has been instrumental throughout this incredible journey. Her creative insights, technical skills and keen eye for detail have elevated the quality and depth of this work.

To my amazing daughter, Eleanor Astras, for inspiring me to look to the future.

To Professor Peter Keil for starting me on this academic voyage. His mentorship has been invaluable, and his friendship has been a constant source of strength and encouragement.

To Harry Fong for repeatedly reminding me to "just write."

To the many other friends, family members, mentors and coworkers who are too many to mention individually.

CONTENTS

PREFACE

As a child, my visits to Lake George were the highlights of my summers. The tranquility of the lake, nestled amid charming homes and bustling businesses, coupled with the endless fun offered by the arcades, ice cream shops and pizzerias, made it a unique vacation destination that felt worlds apart from my home in Brooklyn, New York. Despite being only a few hours away, the experience of Lake George transported me to an entirely different world. Its affordability made it accessible to my family, and we cherished every moment spent there.

Years later, while teaching in New Jersey, I stumbled upon Lake Hopatcong, another gem of a lake that captured my heart. Unlike Lake George, which is surrounded by rural areas, Lake Hopatcong offered a different experience altogether. It was uniquely situated between rural Sussex County on one side and the more bustling Morris County on the other, providing the best of both worlds. This was where I decided to buy my first house.

Lake Hopatcong became my go-to place for quiet and peaceful moments, strolling along the lake and enjoying the stunning scenery. Yet it was never isolated from the world, with Ledgewood's Walmart and Morristown's vibrant social scene just a short drive away. For me, Lake Hopatcong was a hidden gem that offered the perfect blend of serenity and accessibility.

During my pursuit of a doctoral degree in English, I found myself grappling with a difficult question: what should I choose as my dissertation subject? Ideas were swirling around in my head, but none of them seemed to resonate with me. However, as fate would have it, a sudden moment of inspiration struck me while I was enjoying a peaceful lunch on the waterfront.

It was as if a mysterious lake creature had emerged from the depths of the lake, attacking a boat on a foggy, unphotogenic day (as is always the case) and delivering the inspiration I so desperately needed. In that moment, it became clear to me that I should delve into the subject of Lake Hopatcong—not just as a mere object that I was observing in the present but as a critical piece of the historical landscape.

I embarked on a journey of discovery, uncovering the stories of the past that were embedded within the very waters of Lake Hopatcong. As I dug deeper into the lake's history, I began to see it as more than just a picturesque body of water. It was a symbol of the past, a reflection of the cultural and social changes that had shaped the region over the centuries.

Studying Lake Hopatcong allowed me to gain a deeper understanding of the landscape and the important role that it played in shaping the environment, culture and history of the area. Through my research, I was able to gain insight into the ways in which humans had interacted with the lake over time and the impact that those interactions had on the local environs.

Using Lake Hopatcong as the focus for my dissertation was a transformative experience, one that I continue with the writing of this book. It allowed me to see the lake not only as a beautiful and peaceful destination, but also as a window into the past, a critical component of the cultural and environmental landscape. The lessons I learned while studying Lake Hopatcong will stay with me for a lifetime, and I am grateful for the opportunity to have delved into this fascinating subject.

Understanding Lake Hopatcong entails recognizing the intricate layers that have contributed to its development. This book aims to present a comprehensive history of Lake Hopatcong while offering practical insights along the way. It delves into various aspects that shaped the region, including the natural environment, human societies and diverse perspectives. By exploring these different dimensions, we can gain a deeper understanding that enriches our interpretation of the past and enhances our perception of the present and future.

ACKNOWLEDGEMENTS

I wish to extend my appreciation to the Native American communities who originally inhabited the land surrounding Lake Hopatcong and served as its earliest stewards. The profound historical importance of the Lenape people in this region warrants our enduring respect and recognition.

Additionally, I wish to acknowledge the scholars and researchers who have dedicated themselves to the study and documentation of Lake Hopatcong's history. Your academic endeavors have significantly enriched our understanding of this remarkable lake.

I extend my acknowledgement to all those actively involved in supporting and caring for Lake Hopatcong, including the dedicated volunteers.

Lastly, I express my gratitude to The History Press and J. Banks Smither for their invaluable support in making this book a reality.

1
FROM PREHISTORY TO THE COLONIAL ERA

Lake Hopatcong is mostly familiar to those who live in New Jersey and its surrounding areas, such as Manhattan, which is just forty miles away. Currently, it is approximately 4,360 acres in size and reaches a maximum depth of 50 feet. It is also the largest lake in a state known for its famous beaches. Although it stands somewhat in the shadow of the Jersey Shore, it is still popular in its own right for its swimming, fishing and boating. It is also home to popular restaurants and other businesses. Lake Hopatcong is a significant part of the local economy and the natural landscape of New Jersey.

As you venture through the area encompassing Lake Hopatcong, you will come across tangible reminders of its glacial history, serving as evidence of the remarkable geological forces that sculpted the terrain. The undulating contours of the land seem to encapsulate the essence of time and nature, creating a captivating backdrop. Lake Hopatcong's formation and subsequent transformation have been influenced by a multitude of factors over an extensive span of time.

The complex geological history of the region, including the impact of multiple glacial periods, has significantly influenced the landscape of the lake and its surrounding environment. The Pleistocene epoch (2.6 million to 1.8 million years ago) saw the convergence and recession of at least three continental ice sheets across the New Jersey area, with the ice sheets reaching a staggering ten thousand feet thick at their zenith. The effects of these massive ice sheets are still visible today, with many regions in the state bearing the geological traces of the glacial periods.

Lake Hopatcong, dock view. *Author's collection.*

Many of New Jersey's rivers and lakes were formed because of the retreat of these ice sheets. For instance, Peter O. Wacker in *Land and People: A Cultural Geography of Preindustrial New Jersey: Origins and Settlement Patterns* describes, "Glaciation severely altered the surface of the Highlands north of the Wisconsin moraine, scouring upland surfaces and filling valleys with glacial debris." As a result of the formation of these valleys and subsequent sediment deposition, the hills and ridges surrounding the lake were carved out, creating a unique and distinctive landscape.

New Jersey's landscape has also been shaped by volcanic activity. A volcanic eruption roughly 200 million years ago formed the Ramapo Mountains in Northern New Jersey, which are composed of layers of basalt and other volcanic rocks. New Jersey's northern region contains mountain ranges formed by volcanic activity. Furthermore, the region can be split into multiple sections, each oriented toward the northeast and defined by distinct fault lines. Within each section, we can map a geological sequence of parageneses, alongside granitic or syenitic rocks.

As a result of past volcanic activity, the Lake Hopatcong area has many different types of rocks, the main one being a gneiss made up of quartz,

Nepheline syenite rock. *From Wikimedia Commons.*

microcline microperthite, plagioclase and biotite. There are also other types of gneiss present, such as amphibolite and hornblende-quartz-feldspar gneiss. The layers of rock can vary in thickness from 0.5 to 4.0 inches, and the entire units can range in thickness from 100 to 1,500 feet. About 40 percent of the rocks in the area are biotite-feldspar-quartz gneisses, which are believed to have come from sandstone parent rocks with high potassium levels. These types of rocks are also common in other areas in the Reading Prong. Due to Lake Hopatcong's granite-laden granite layer, its geological characteristics remain an obstacle to modern development. These features serve as a reminder of the past.

Fast forward, and Lake Hopatcong now sprawls over 2,658 acres, straddling the border between Morris and Sussex Counties, its waters reaching a maximum depth of about 50 feet. Lake Hopatcong is a vibrant echo of history and plays an instrumental role in New Jersey's current ecological landscape. This disproportionate impact comes down to lakes' superior capacity to sequester carbon, thanks to their relatively high levels of carbon dioxide and methane gas. Such active retention of carbon dioxide by lakes helps curtail the presence of greenhouse gasses in our atmosphere. Lakes also perform other remarkable environmental feats. They regulate temperatures and function as nature's sponges, absorbing rainfall to curtail soil erosion and flooding. Their ability to temper the severity of extreme weather events also testifies to their crucial role in our planet's wellbeing.

Lake Hopatcong's Features

Lake Hopatcong's inherent charm is accentuated by the Musconetcong River, a vital lifeline that courses through the landscape before ending its northward journey in the lake. This river is not just a water source; it's the lifeblood of a myriad of plants and animals that call the area home. Stretching around fifty miles long, the Musconetcong River is an enchanting trail of sparkling water that weaves its way through diverse terrains. It breathes life into the dense forests, imbues farmlands with vitality and carves a shimmering path through bustling urban centers before uniting with the majestic Delaware River in Pennsylvania.

The lake serves as a lively and bustling habitat, teeming with a diverse assortment of life. It's a vibrant underwater world brimming with a spectrum of aquatic creatures. Among them, nimble fish dart through the waters, weaving intricate patterns beneath the surface. In contrast, turtles move at a more measured pace, drifting slowly and peacefully into the depths.

Opposite: The Musconetcong River. *Author's collection.*

Above: A duck on Lake Hopatcong. *Author's collection.*

Left: Hibiscus at Lake Hopatcong. *Author's collection.*

Above the water, the lake offers a spectacle for birdwatchers. Majestic herons stand tall on their long legs, scanning the waters. The peaceful duck meanders across the lake surface. Meanwhile, fish, such as largemouth bass, muskellunge and bluegill, create little splashes that dot the tranquil lake. Now and then, mighty eagles make guest appearances. Their flights, like a ballet in the sky, paint captivating silhouettes against the shimmering lake surface. All this wildlife reminds us of nature's wonder and beauty.

Flora at Lake Hopatcong is equally impressive, boasting a plethora of plants that line its banks and dive beneath its waves. From luscious water lilies and delicate reeds to towering oaks and maples, the greenery around the lake paints a vibrant picture of Mother Nature's artistry. Lake Hopatcong, with its magnetic charm, captures the heart of nature lovers, adventurers and casual visitors. It's a place where people can unwind to the soothing lullabies of lapping waves, lose themselves in the mesmerizing hum of insects and marvel at the astounding vistas unfolding before their eyes. Truly, the beauty of Lake Hopatcong and its river companion, the Musconetcong River, offer a soul-stirring ode to the wonders of the natural world.

Early Humans: The Lenape and Paleo-Indians

Estimates suggest that the first humans appeared on Earth around 15,000 to 12,000 BCE. During this ancient era, one geographical feature played an especially significant role in human migration: the Bering Land Bridge. This natural bridge is one of the most well-known in history, and it emerged during the last ice age, connecting Asia and North America. The Bering Land Bridge provided a passageway for humans and animals, acting like a highway between the two continents. This land connection allowed species to migrate across the icy expanse, introducing new life to unfamiliar lands. Among these travelers were our human ancestors, who crossed this bridge to reach the Americas and settle there. The role of the Bering Land Bridge in human migration and animal dispersal was pivotal. The bridge essentially opened a door, allowing travel between two vast continents.

The Lenape, a Native tribe that settled in what is currently known as New Jersey, arrived approximately three millennia ago. Prior to their arrival, it is speculated that another group of individuals whose tribal affiliation remains unknown inhabited the region. Even so, there are only marginal differences in cultural practices between the last prehistoric epoch and the initial historic period based on archaeological records.

In all likelihood, Northern New Jersey was inhabited by Paleo-Indians before the Lenape arrived, paving the way for later groups to settle here. However, the nature of the relationship between the Lenape and the Paleo-Indians remains uncertain.

A further layer of complexity arises from the fact that the terms *Paleo-Indian* and *Lenape* are often used interchangeably. It is likely, however, that the Lenape and Paleo-Indians had unique cultural and societal customs. In the future, archaeological evidence and further studies may provide a better understanding of the interplay between these groups. By unraveling the distinct influences and interactions they had with each other, we will gain valuable insights into how they shaped the course of local history. These explorations will not only expand our comprehension of the past but also illuminate the trajectory of human societies in this region.

Evidence of the Past

Understanding the complex lives and cultural practices of the early inhabitants of any region remains an arduous task, particularly due to the lack of comprehensive historical records from these societies. However, it is important to note that the gaps in our knowledge do not signify the absence of relevant information. Through the meticulous examination of artifacts and other physical remains, a significant amount of information has been uncovered. One such source of insight is Herbert C. Kraft's book *The Lenape-Delaware Indian Heritage: 10,000 BC to AD 2000*, wherein he provides thorough descriptions of the early people based on environmental discoveries at excavation sites. Notably, Kraft's excavation at the Plenge site, which is situated approximately twenty miles from present-day Washington, New Jersey, yielded valuable information.

In the broad landscape of academic research, Joseph A.M. Gingerich's book *In the Eastern Fluted Point Tradition* holds a distinguished place as a comprehensive study on the early inhabitants of New Jersey. This influential work dissects the history and culture of these peoples with remarkable detail, focusing notably on the largest known Paleoindian site in the state, situated alongside the Musconetcong River in Northwestern New Jersey. His work dives deep into the complexities of the Paleo-Indian society of the region. Gingerich's scholarly analysis furnishes an intricate portrayal of their daily life, yielding insights into a range of cultural aspects. From domestic affairs to subsistence strategies—and from their material culture practices to their societal structures—Gingerich's exploration paints a vivid picture of this early society.

The meticulous analysis of archaeological artifacts excavated at the Plenge site has provided scholars with a profound comprehension of the abundant historical significance pertaining to Lake Hopatcong and its environs. The Plenge site stands out among other sites due to its numerous distinct arrowhead designs, which are rarely found elsewhere. This discovery provides strong evidence suggesting that early communities occupied this area for an extended period as part of their broader settlement patterns. This underlines the Plenge site's pivotal role as an invaluable repository of prehistoric heritage in the region. Consequently, it stands as an indispensable asset for researchers and plays a critical role in expanding our knowledge of the area's past.

Distinguished anthropologist Ronald J. Mason discerns that this era bore witness "to widespread physical and biological transformations in the world about him—the biological changes being of a magnitude unprecedented in Cenozoic time." This statement underscores the exceptional nature of the biological modifications that took place during the time of early humans. These transformations serve as a critical focal point for understanding the significant shifts in the living organisms and ecosystems of the past, as well as the unique circumstances and challenges faced by our ancestors. By recognizing the unprecedented magnitude of these biological changes, we gain insights into the impact and influence early human communities had on their environment, as well as the evolutionary dynamics that shaped both human and nonhuman life during that period.

Late Pleistocene-Colonial Time Periods

As anthropologist Ronald J. Mason alluded, a series of significant events and transformations have transpired throughout human existence. By delving into a comprehensive understanding of these chronological epochs, we can cultivate a heightened appreciation for this intricate and captivating history, as well as understand how it has shaped our contemporary world, including Lake Hopatcong. This section will provide a timeline from the late Pleistocene to colonial periods. This timeline of events has led to the creation of the unique and diverse cultural and environmental landscape that we see in New Jersey today.

The late Pleistocene period (10,000 BCE to 8,000 BCE), also called the Paleo-Indian period, was a time of early people. Like the rest of the world, New Jersey went through significant changes during this time, including the extinction of many large mammal species. Before 10,000 BCE, Lake Hopatcong was hidden under the Wisconsin Glacier, which lasted from 21,000 BCE to 13,000 BCE. However, as the glaciers began to recede, a process of glacial melting commenced, giving rise to the proliferation of lush vegetation that enticed early human settlements. This remarkable period is notably distinguished by the unearthing of predominantly stone and clay artifacts. Although the passage of time has obscured many aspects of this early chapter of human history, astute archaeologists adeptly glean essential insights from meticulously examining the geological characteristics of New Jersey and scrutinizing the diverse array of artifacts and other remnants in this state that originated from this epoch.

During this period, the climate was characterized by Kraft as "cool and dry," and humans relied on hunting, fishing and gathering as their main sources of sustenance. They inhabited diverse landscapes, such as park tundras, forests and grasslands, and utilized a variety of tools, such as fluted spearpoints, knives and scrapers. Through detailed analysis of pollen and carbonized seeds discovered in charred remains from hearths, researchers have uncovered a wide range of edible seeds and plants. Also according to Kraft, these include "lamb's quarters, goosefoot, ground cherry, blackberry, hawthorn plum, pokeweed, pigweed, smartweed, lettuce, grape, hackberry and sedge." The presence of such a diverse array of plants indicates a sophisticated understanding of agriculture and a deep knowledge of the local environment.

In contrast to previous assumptions, evidence suggests that individuals may have established settlements instead of continually moving. They would relocate only when resources became depleted. This insight provides information about the lifestyles and subsistence practices of early inhabitants, as well as the ecological conditions and available resources of the time. It suggests that these individuals had a profound understanding of the environment and utilized its resources effectively. This ability to develop long-term strategies for subsistence contributed to these early people's overall success in adapting to and thriving in their surroundings.

Meanwhile, during this formative period of Lake Hopatcong's evolution, the basin did not resemble the present-day lake in its entirety. It is likely that the area was enshrouded in ice, punctuated by water bodies that provided fresh water and fishing opportunities. In spite of their short tenure, the inhabitants of the area managed to stabilize their living conditions. The types of animals that the Native hunters pursued remain uncertain, with some experts opining that they may have gone after larger creatures like mammoths and mastodons. The assertion is supported by artifacts found at other hunting sites, such as those in Kenosha County, Kansas. Early settlers might have cultivated the land for a time before moving to areas with greater hunting opportunities. These findings help shed light on the Native hunters' lifestyles and their impacts on the environment.

Additional resources based on excavations in and around the Plenge site support that the Paleo-Indians during the late Pleistocene period also lived alongside mastodons and other large, now-extinct creatures. Mastodon relics discovered near present-day Vernon, New Jersey, were subjected to carbon dating, resulting in an age of approximately 10,890 ± 200 BP. Experts theorize that mastodons became extinct because Paleo-Indians overhunted them. In other states, such as Ohio, there is evidence of Paleo-Indians cooking and eating mastodons, but archaeologists have found no such evidence in New Jersey, as no fluted points were found with or near any mastodon remains. Determining whether mastodons were hunted and consumed is challenging due to the combination of increasing sea levels and previous careless excavations.

Moreover, the rise in sea levels during the early Holocene period (circa 8,000 BCE to 6,500 BCE), also known as the early Archaic period, likely contributed to the formation of Lake Hopatcong. The climate around Lake Hopatcong at that time would have been warmer and wetter than it was during the late Pleistocene period. For the most part, human tools remained relatively unchanged. The warmer, dry climate produced boreal forests. A variety of plants would have been found around the region. These included some plants that had adapted to the changing climate and some that were more suited to the colder climate during the late Pleistocene. Humans would have been able to adapt to changing environmental conditions.

In the middle Holocene period (circa 6,500 BCE to 4,000 BCE), also known as the middle Archaic period, the climate turned "warm and wet," leading to "mixed forests of mostly hemlock and oak." Kraft adds:

> *An ameliorating climate, rise in ocean levels, and the spread of mixed forests replaced the megafauna and flora of Paleo-Indian times with an abundance of nut trees and an increase in deer, elk, bear, turkey, and other creatures who thrive on mast foods, leaves, and grasses.*

An exhibit in the geology museum at Rutgers University in New Brunswick, New Jersey. *From Wikimedia Commons.*

While rising sea levels have negative consequences for our planet today, during this period, they led to advantageous conditions for the quality of human life, since "springs and lakes provided potable water, game, edible plant foods, and firewood."

From approximately 4,000 BCE to 2,000 BCE, the late-middle Holocene period, also known as the late Archaic period, climatic conditions became what Kraft calls "warm and dry." Humans developed more sophisticated tools, including fish nets, axes, knives and baskets, which allowed them to utilize the environment more effectively. This technological advancement led to a marked increase in gathering activities, as it facilitated the exploitation of various food sources in the region, including a rich mix of oak, hickory and chestnut trees.

Furthermore, during the late-middle Holocene period, the availability of favorable food resources and the development of more sophisticated tools may have also influenced regional trade and exchange networks. It is likely that the ability to produce specialized goods and services, such as high-quality baskets or expertly crafted tools, contributed to an increase in social interactions and an exchange of goods and ideas across wider geographic regions. In general, the late-middle Holocene period was pivotal in the history of the region, characterized by significant changes in material culture, subsistence practices and social structures.

In the late Holocene period, which spans from approximately 2,000 BCE to 1,700 CE, there were five distinct periods, each marked by unique environmental and cultural changes. From around 2,000 BCE to 1,000 BCE, the Terminal Archaic period was characterized by a continuation of the warm, dry conditions of the preceding late Archaic period. During this time, the region's forests became more mixed, and Natives developed an array of tools that facilitated settlement and subsistence. These tools included scrapers, soapstone bowls and kettles that were used to cook and prepare food.

Because soapstone kettles took time and effort to make, they played a crucial role in settlement development. People invested in a more stable and permanent living environment by creating these kettles, which were flat with handles used for holding them over a fire. The region rapidly expanded as people settled and began further cultivating their land. After the development of cooking utensils, it was possible for Natives to prepare food more efficiently, maximizing resources. In turn, this enabled a growing population to live and thrive in the region for centuries to come.

During the early Woodland period, which spanned from around 1,000 BCE to 0 CE, New Jersey's landscape predominantly featured dense forests

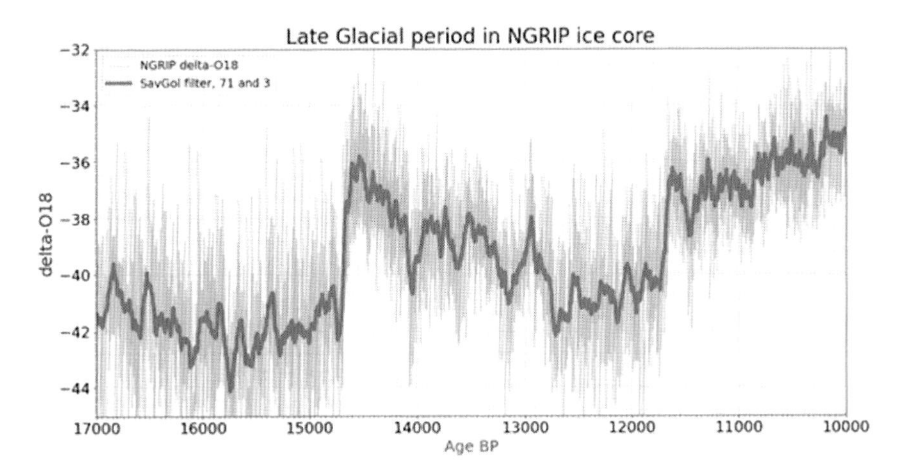

Warming in the late Holocene period. *From Wikimedia Commons.*

of oak and chestnut trees. This era was marked by significant advancements in pottery craftsmanship, leading to an upsurge in pottery production and subsequently fostering increased trade interactions among different groups.

As a response to this newfound stability and growing trade networks, campsites expanded in size, and people began to settle in one place for more extended periods. During this time, the general living conditions involved the use of bark or thatch huts, as well as various tools and technologies such as fireplaces, dugout canoes, grooved axes, soapstone pots and ceramic vessels. Certain areas within New Jersey were favored for human settlement, including terraces overlooking large streams, marshlands and coastal sites. These locations offered advantageous conditions for habitation and resource utilization during the early Woodland period.

This period is particularly intriguing for archaeologists and historians, as it witnessed noteworthy changes in the types of artifacts crafted by people. During this era, a significant development was the initiation of ritual burials. Notably, in Summit, New Jersey, subrectangular gorgets featuring engraved human stick figures have been unearthed. Although there's a possibility that these artifacts were used in burial practices, the extent of their usage remains uncertain, given the absence of similar gorgets in other known contexts. The presence of ritual burials serves as a significant indicator of an increasing cultural complexity during this period.

The middle Woodland period (circa 1 CE to 1,000 CE) is significant in the history of the region now known as Lenapehoking. It was during this period that the Lenape people arrived, and they brought advanced hunting techniques and tools, including bows and arrows. At this time, the landscape of the region underwent significant changes, resulting in what Kraft describes as "modern forests and grassy vegetation." Amid these transformations, the Lenape people established a principal settlement near Kipp Island, an inlet in northeastern Pennsylvania. Notably, the site has over two hundred ceremonial burial sites that were left by the Nariticong, a branch of the Lenni Lenape. This rich historical site reveals the depth of cultural practices that were already developing.

Furthermore, the Lenape people were on the cusp of developing agriculture during this time, as evidenced by their cultivation of tobacco. Despite the popular belief that the Lenape were primarily hunters and gatherers, these agricultural pursuits demonstrate their increasing mastery over the land and their ability to sustain themselves. Moreover, the discovery of cylindrical pestles in the region provides further evidence of their sophisticated agricultural techniques, specifically in maize preparation. Based on archaeological records, the Lenape people were, on average, approximately 1.7 meters tall during this period and wore animal skins and feather robes. It was easy for the Lenape people to utilize the resources around them efficiently, as these clothing materials were readily available in the region. The middle Woodland period was a time of significant change and growth for the Lenape people in Lenapehoking, as they developed new survival techniques and adjusted to the changing landscape.

The Lenape faced another wave of development during the late Woodland period when the climate conditions reached an approximation of the present-day climate. During this period (approximately 1,000 CE to 1,650 CE), tool complexity increased and agriculture became more prevalent. A number of bone tools, like needles and awls, indicate the development of sewing, while antler tools, such as arrow points, scrapers and flakers, indicate advancements in stone artifacts. The Lenape also started to cultivate crops during this time, drying and storing fruits for winter. Hunting became less important, and agriculture was likely to become the primary source of food for the community.

During this period, an increase in population intensified competition for resources and gave rise to conflicts. This marked a significant transformation in Native societies and economies, as communities shifted from living in dispersed homesteads to forming larger settlements with clusters of houses, ultimately progressing toward the establishment of fortified villages. These changes reflect the dynamic nature of Native life during this era and the strategies employed to adapt to evolving social and environmental changes.

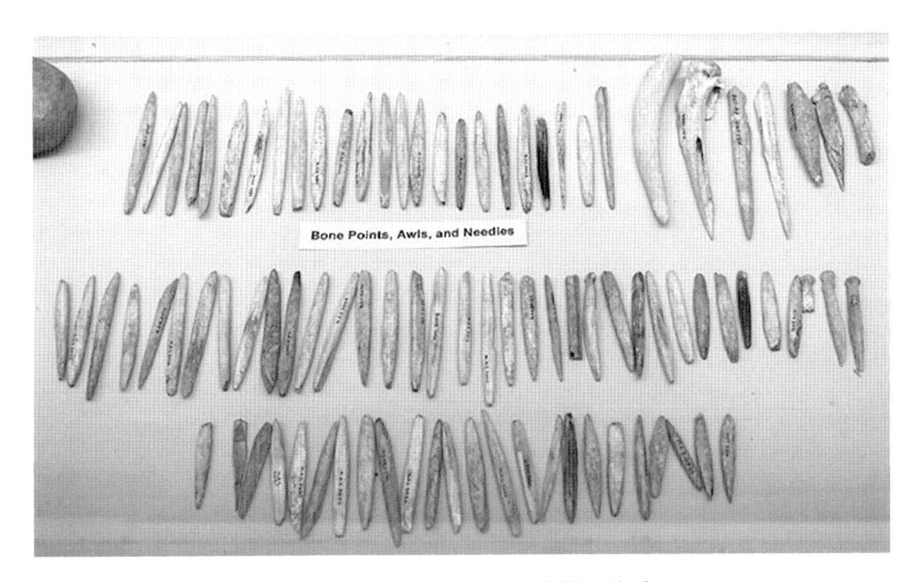

Needles and awls from the late woodland period. *From Wikimedia Commons.*

In order to efficiently function, everyone played a role in sustaining each community. Men burned and cleared fields for cultivation, fertilizing them with ash to sustain the community's livelihoods. The Lenape communities developed distinct practices. For instance, although fish had been used as fertilizer in New England, no evidence of this practice among the Lenapehoking people has been found. The Lenape maintained soil fertility by moving their cultivation fields frequently, with each family taking care of approximately two to three acres. The community's food production was also largely dependent on women and children, who gathered wild plants such as berries and herbs. In adapting to the modern environment, the Lenape people demonstrated their resilience and ability to thrive in changing circumstances. As a result of using new tools and developing agricultural practices, the Lenapehoking community was able to sustain itself and thrive.

Nonetheless, things did change. The historic contact period, spanning approximately from 1,600 to 1,700 CE, ushered in a profound transformation for the Lenape people. This period marked their initial encounters with colonial ships navigating the Delaware River, signifying a momentous turning point in their history that would have enduring consequences. The arrival of foreign material goods and people brought about an irreversible alteration in Lenape culture and their dynamic with the surrounding environment. This transformation is exemplified by their adoption of iron tools and firearms. Furthermore, interactions between

Above: A display showing Lenni Lenape Natives at Paterson Museum in Paterson, New Jersey. *From Wikimedia Commons*.

Opposite: Chalk portrait: the only known image of William Penn's likeness known to have been made during his life. *From Wikimedia Commons.*

the Lenape and Europeans introduced them to a diverse array of novel materials, including kaolin pipes and metal objects. These innovative tools exerted a considerable influence on the Lenape's hunting and agricultural practices, fundamentally reshaping their way of life.

From European records, we gain our first textual descriptions of the Lenape during this period. The Lenape were described as "well proportioned, slender" and "brown-skinned with yellowish or reddish hues." Their "incisors, especially the middle ones, were often shovel-shaped, a Mongoloid characteristic." European accounts also praised the Lenape's strengths and skills, noting their acute "senses of hearing, sight and smell," as well as their abilities with their hands. Additionally, the Lenape were described as possessing "a good memory, lively imagination, genuine wit, natural understanding" and a strong sense of curiosity. These observations provide valuable insights into the Lenape's physical characteristics, social dynamics and cultural practices during this period.

A notable historical account that discusses the Lenape during the colonial period is that of William Penn (1644–1718), an English Quaker who founded

the Province of Pennsylvania (now Delaware, Pennsylvania and New Jersey) in 1683. In his writings, provided by archaeologist and author Dorothy Cross, Penn provides a detailed description of the physical attributes of the Lenape people:

> *For their Persons, they are generally tall, streight* [sic], *well-built, and of singular proportion; they tread strong and clever, and mostly walk with a lofty Chin: of Complexion, Black, but by design, as the Gypsies in England: They grease themselves with Bears-fat clarified, and using no defence* [sic] *against Sun or Weather, their skins must needs be swarthy.*

Despite Penn's writings being from a European perspective, his account provides insights into the physical characteristics of the Lenape people.

Moreover, unlike others, Penn refrains from making any significant personal commentary on the Lenape's appearance, which demonstrates his desire to describe them objectively. As such, it can be concluded that the Lenape were in good physical condition.

In addition to noting other facts, Penn described the language of the Native population. Penn's impressions of the Lenape language are as follows:

> *Their Language is lofty, yet narrow, but like the Hebrew; in Significance full, like Short-hand in writing; one word serveth in the place of three, and the rest are supplied by the Understanding of the Hearer: Imperfect in their Tenses, wanting in their Moods, Participles, Adverbs, Conjunctions, Interjections: I have made it my business to understand it, that I might not want an Interpreter on any occasion: And I must say, that I know not a Language spoken in Europe, that hath words of more sweetness or greatness, in Accent and Emphasis, than theirs.*

Unlike later accounts from other explorers, the comment regarding "sweetness" suggests some appreciation for the local culture.

INFUSION OF TWO CULTURES

Colonists soon realized that specific facets of the natural surroundings bore the unmistakable marks of cultural influence. In 1715, John Reading, an English Quaker born in Gloucester, New Jersey, offered valuable insights into the region. He had been sent to England for his education, and after spending several years there, he returned to America to join his father in managing their extensive landholdings. Following his father's death in 1717, Reading inherited a substantial estate primarily consisting of land, which positioned him as the wealthiest individual in Hunterdon County. Notably, he served as the governor of New Jersey in 1747 and held the position once again from September 1757 to June 1758.

John Reading's explorations extended to Northern New Jersey, where he sought resources, habitable land and Native settlements that could be utilized by Europeans. In his observations, Reading found both Morris and Sussex to be characterized by rocky and mountainous terrain, limiting their potential for anything other than iron and timber extraction. While he noticed the region's abundance of fish, rendering it visually appealing for recreational purposes, his party faced numerous challenges while crossing the rugged landscape. He encountered arduous climbs, encounters with rattlesnakes, persistent mosquitoes and gnats and intense thunderstorms. Exploring remote areas meant limited access to essential supplies, including food, water and shelter. Unlike modern explorers with advanced GPS technology, Reading had limited tools for accurate mapping and navigation, making it easier to get lost or disoriented.

However, John Reading's accounts of the Musconetcong River region proved invaluable for gaining insights into the practices within the area. Of particular note was an extensive plain he described as "barren," a direct result of the Lenape's deliberate use of controlled fires to clear the land. This practice, deeply rooted in cultural choices, left a lasting imprint on the land. These intentional fires created open spaces, significantly altering the area's ecological composition. Recent botanical research indicates that oak trees, rather than pine forests, were originally the dominant natural vegetation in the region. Pine forests typically develop when fires burn an area at regular intervals for ten to twenty years. Despite the advantages of burning vegetation, Reading was critical of the impact these fires had caused, stating, "This advantage does not compensate for the harm done by the fires." The damage was substantial, but it did create a suitable living environment for the Lenape people and even facilitated the domestication of livestock in the region. While some species declined due to these fires, others adapted and thrived. Ultimately, the Lenape people's practice of controlled burning left an enduring and profound mark on the region's biological composition that would influence the interactions of future generations with the land.

These fires also served other essential needs for the Lenape. They used the smoke to communicate, enabling them to signal camp locations to neighboring tribes. These carefully orchestrated fires were also used to drive out game, facilitate the movement of animals such as hogs, improve travel conditions and enhance visibility. Although earlier descriptions of the land characterized it as "virgin," devoid of both human presence and design, contemporary research shows otherwise.

European explorers did not merely observe the Lenape landscape; their arrival fundamentally reshaped it. To comprehend the profound impact of this encounter, it is essential to delve into the intricate relationship between the Lenape people and the natural world, as it provides critical insights into how this alteration disrupted their way of life. While the Lenape shaped their environment, they always strove to maintain a balance because central to Lenape's worldview was the concept of *manëtu*, or the idea that all objects contain a spirit. For them, the natural world was infused with spirits, whether they came from animals, plants or other elements of the environment. In their cosmology, all living beings and natural features were interconnected and played an important role in maintaining harmony within the environment.

Not all spirits were benevolent; some held the potential to bring harm, serving as essential components in maintaining a balance within the spiritual world. The Natives recognized that even seemingly minor natural elements, like a stone or a gust of wind, could be influenced by damaging spirits in some unknown way. To mitigate the risks associated with encountering these potentially hazardous elements, Natives strived to coexist harmoniously with the natural world, maintaining a delicate equilibrium. One of their main strategies to achieve this goal was the sustainable use of natural resources, and they were careful never to take more than they needed. By passing these values down from generation to generation, they ensured that respect for nature endured for centuries and became an integral part of their culture.

In contemporary times, the concept of *manëtu* retains its relevance, highlighting the significance of living in concord with nature and recognizing the impact of human activities on the environment. Many Native cultures continue to uphold similar beliefs, which have played a significant role in shaping modern perspectives on environmental sustainability and environmentalism.

Maintaining a balance is difficult. Unintentional but excessive resource consumption can yield unforeseen and unpredictable consequences. This is exemplified by the mastodon's becoming extinct because of overhunting. This disruption had profound repercussions for the many Native communities, compelling them to adapt their way of life and seek alternative sources of sustenance in a continually changing environment. Conversely, responsible cultivation practices enabled the production of ample quantities of corn and other crops to meet the people's dietary requirements. However, even with meticulous management, such as altering the ecosystem could still result in unforeseen and potentially adverse effects. If the *manëtu* is displeased, individuals must either seek reconciliation or face the repercussions, which may manifest as unfavorable weather conditions or other adverse events.

The arrival of Europeans led to rapid changes disrupting all existing harmony. It is undeniable that colonization led to the largest topological changes in the region. Wacker writes:

> *It is also necessary to keep in mind that certain aspects of the physical environment had been culturally induced well before the time of first European contact. Aboriginal practices, both conscious and unconscious, had induced a plant and animal cover much more agreeable to man than would have been the case under purely natural circumstances. In fact, much of the favor with which Europeans initially viewed the area, as well as many of the subsequent activities, relied upon the fact that the environment had been altered.*

Native people developed strong navigational skills to facilitate hunting and gathering across diverse landscapes. Through the use of tools and resources, they cultivated sustainable community practices that would later prove useful to colonial explorers and settlers.

At the outset, the Lenape experienced relatively peaceful interactions with Europeans, engaging in the shared use of land and the exchange of knowledge and resources, notably through fur and goods trading. However, as the region witnessed a growing influx of settlers, tensions escalated, resulting in a breach of their land ethics and, ultimately, their forced displacement. Tragically, many Lenape fell victim to diseases introduced by Europeans, while others lost their lives in violent conflicts. These developments marked the beginning of their gradual displacement from their ancestral lands, a topic that will be further explored in chapter 2 of this book.

Embracing the Past in the Present

The lasting influence of the Lenape people is discernible in the present-day development of roads and local nomenclature in the vicinity of Lake Hopatcong. Through their land-clearing efforts, the Lenape paved the way for the construction of numerous contemporary roads, many of which still retain Native names. Amid this collection of modern settlements, where a vibrant gas station stands on Hopatchung Road, spirited crowds gather to support the high school's Hopatcong Chiefs and motorboats glide gracefully across the lake, the echoes of the past resonate. However, conspicuously absent is a tribal area, serving as a poignant reminder of the profound historical impact when considering the lake's broader historical context.

Numerous trails once employed by the Lenape as a practical transportation network have evolved into significant roadways. One prominent example is Route 10, a pivotal route facilitating travel in and out of Lake Hopatcong.

Additionally, the design and construction patterns of present-day Lake Hopatcong still bear the hallmarks of Lenape influence. The Lenape established their communities in proximity to streams and wetlands, ensuring easy access to food resources and transportation. Subsequent settlers, following in their footsteps, perpetuated this tradition, ultimately shaping the contemporary development of Lake Hopatcong.

2
HISTORIC CONTACT AND CULTURAL SHIFTS IN THE SEVENTEENTH AND EARLY EIGHTEENTH CENTURIES

From the time when the first humans set foot in the Western Hemisphere and discovered Lake Hopatcong, this region has maintained an enduring connection with people. The lake has consistently played a vital role in providing sustenance and serving as a source of leisure activities for countless generations. Its breathtaking natural beauty has been a wellspring of inspiration for countless artistic endeavors, spanning visual arts, poetic compositions and literary works. As a result, Lake Hopatcong has earned its place as a beloved landmark, cherished by both the local community and visitors from around the world.

Today, a statue paying homage to the Lenape tribe—yet bearing no name—stands majestically in the middle of Maxim Glen Park, located in Hopatcong. The monument artistically depicts a scene from a bygone era: a man firmly seated on a horse's back, an untethered horse on his left. His posture and firm grip suggest a calm authority, indicating his adeptness and comfort with the horses. From the vantage point of Hopatchung Road, this tribute is placed strategically on top of a small hill. This elevation adds to the statue's striking presence, amplifying its silent storytelling. The figure and his horses, frozen in mid-motion, appear like a timeless tableau, captured from a distant past yet standing resiliently in the present. It is as if at any moment, they could break from their bronze stillness and begin to roam the park's expansive grounds.

In an arresting contrast, however, the view from behind the statue tells a different tale. Today, the man would be seeing a landscape that is starkly

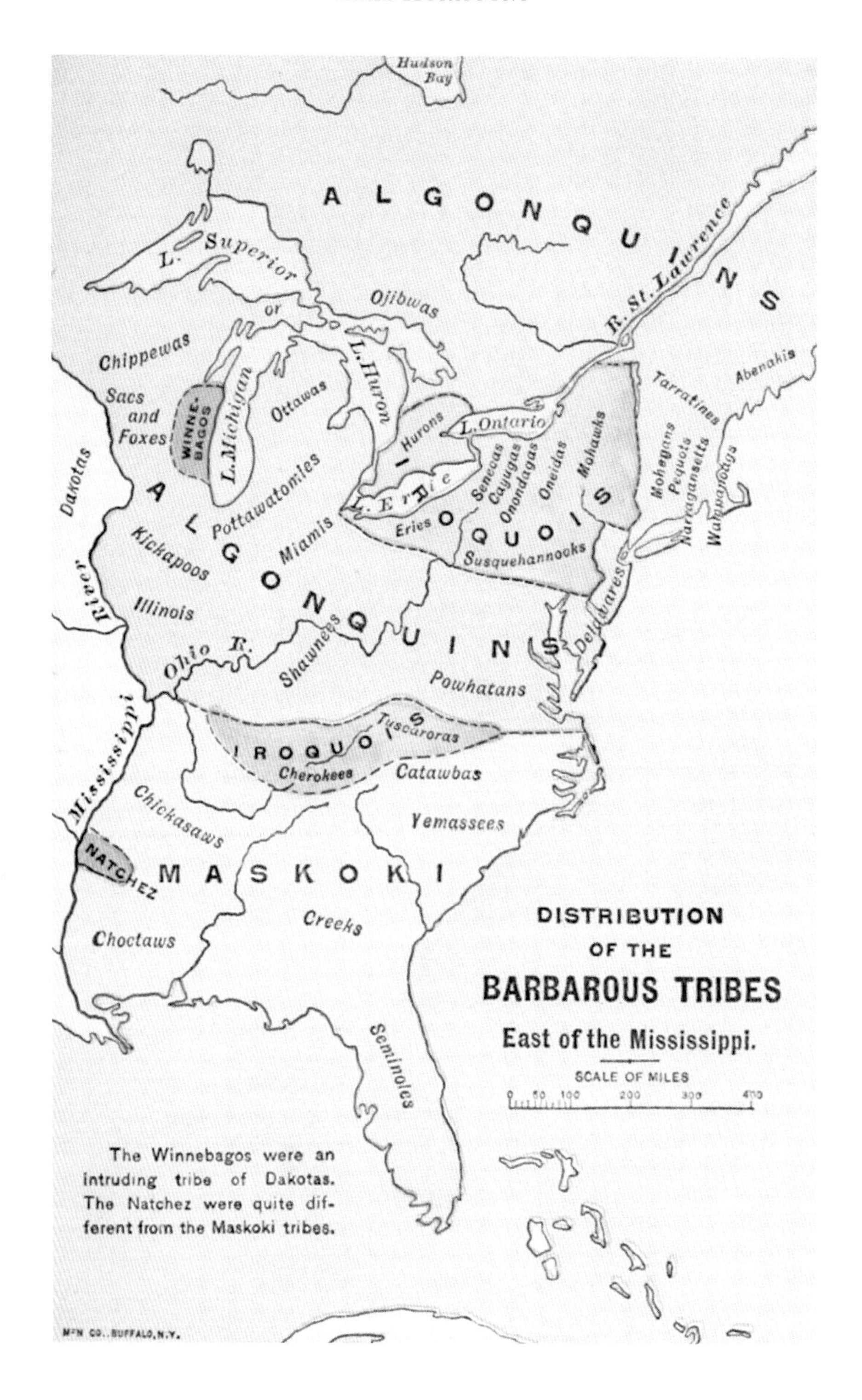
Hudson Bay
ALGONQUINS
L. Superior
Ojibwas
R. St. Lawrence
Chippewas
Sacs and Foxes
WINNEBAGOS
L. Michigan
Ottawas
L. Huron
Hurons
L. Ontario
Abenakis
Tarratines
Dakotas
Pottawatomies
L. Erie
IROQUOIS
Senecas
Cayugas
Onondagas
Oneidas
Mohawks
Mohegans
Pequots
Narragansetts
Wampanoags
Kickapoos
Miamis
Eries
Susquehannooks
River
Illinois
Delawares
Ohio R.
Shawnees
Powhatans
Mississippi
Tuscaroras
Cherokees
Catawbas
Chickasaws
Yemassees
NATCHEZ
MASKOKI
Creeks
Choctaws
DISTRIBUTION
OF THE
BARBAROUS TRIBES
East of the Mississippi.
SCALE OF MILES
0 50 100 200 300 400
Seminoles
The Winnebagos were an intruding tribe of Dakotas. The Natchez were quite different from the Maskoki tribes.
M'F'N CO., BUFFALO, N.Y.

Opposite: From an 1899 textbook called *A History of the United States for Schools*, which shows a biased history of "barbarous tribes." *From Wikimedia Commons.*

This page, top: Looking up at the Lenape statue at Maxim Glen. *Author's collection.*

This page, bottom: A view from behind the Lenape statue at Maxim Glen. *Author's collection.*

different from the land of his time. Instead of open plains and boundless forests, his eyes would meet the trappings of modern civilization: political signs advertising local elections, the changing colors of a traffic light, a constellation of modern houses dotting his sight and other elements of today's world. This anachronistic juxtaposition seems to place the silent man as an ever-vigilant sentinel over his ancestral lands. He becomes an observer of the centuries of change that have swept across his homeland. It truly stirs the imagination to ponder what thoughts might be crossing this man's mind. From his vantage point, we are invited to reflect on the layers of history that make up our present and contemplate the balance between progress and preservation.

CLASHING WORLDS

I attempt to envision what that moment must have been like. In my mind's eye, I see it unfolding: as the Lenape carried on with their everyday activities, an extraordinary sight suddenly materialized on the horizon before them. They witnessed the arrival of foreigners, individuals whose attire was unfamiliar, their speech incomprehensible and their possessions unlike anything the Lenape had ever encountered. Fueled by curiosity and perhaps a touch of caution, the Lenape gathered to observe these enigmatic visitors. Their apprehension was evident as they watched the strangers draw nearer, and an overwhelming sense of uncertainty permeated the atmosphere.

The Lenape could not determine the origins of the strangers or predict their potential actions. There is a chance that they had encountered tales of other societies afflicted by the diseases introduced by European settlers and consequently adopted a cautious stance. As they scrutinized the activities of the newcomers, the Lenape were aware that their world was undergoing a significant transformation. They were coming to terms with the probability that their customary way of life was on the brink of irrevocable change.

On the other side, the early European settlers—mainly Dutch, Swedish and English—were more focused on their own survival and prosperity. In their quest for a new life, they often ignored the Native people who already lived on the land. Much of their narratives from this time originate from an array of journals. However, we have only indirect accounts about the Lenape tribe. Many of these accounts were written by missionaries who lived among the Lenape and made extensive notes about their culture. But these reports were often distorted by the missionaries' own cultural perspectives and religious biases. This resulted in a distorted view of Native culture and spirituality, which was usually shown in a negative light. This has led to more stereotypes and wrong ideas about Native identity.

This form of biased storytelling finds its origins in the early contact between Christopher Columbus and the Americas in 1492. Columbus's narratives are well-known for their Eurocentric perspectives when describing Native people. Columbus labeled the Indigenous peoples he encountered as "Indians," a misnomer stemming from his erroneous conviction that he had reached the Indian Ocean, even though he was in the Caribbean. This misidentification endured, and to this day, Indigenous populations in the Americas are at times still alluded to as "Indians." Columbus exhibited little respect for the people he encountered, and his crew subjected many Native Americans to enslavement, coercing them into labor on ships, in mines and on plantations while subjecting them to harsh treatment and exploitation.

Moving forwarding to the 1600s, the arrival of Henry Hudson brought Native populations more prominently into European consciousness. Nevertheless, Hudson's accounts were likewise tainted with bias, featuring stark warnings against placing trust in Native communities and boasting about Europeans' capacity to displace them from their ancestral lands. This dismissive portrayal of Natives and their cultures persisted throughout the colonization period in the Americas. Such derogatory narratives facilitated the justifications used by colonizers to lay claim to lands that had been home to Native peoples for centuries.

Contrasting Perspectives: Dutch and English Impressions

The arrival of early European settlers left an indelible mark on both the Lenape people and the wider region of New Jersey. From the beginning of European contact, Lake Hopatcong held a position of interest for explorers. A map of New Jersey, believed to have been created in 1656 by Dutchman Van der Donk (though there is some debate about its origin), included the lake, then referred to as Huppakong. Interestingly, this map mistakenly connected the lake to the Hudson River, though the error was not much more significant than a later misreport in a New York newspaper that wrongly placed Lake Hopatcong in the Pocono Mountains. Despite these flaws, this map contributes to our understanding of Lake Hopatcong's early significance and its role in the unfolding story of European exploration and colonization in North America.

The Dutch in New Jersey heralded an era characterized by extensive colonialism that would endure for centuries. The establishment of Dutch settlements played a pivotal role in shaping the economic landscape and overall development of the region. The Dutch West India Company, formed in 1621, laid the groundwork for exploration and colonization. The company's primary objectives were to establish colonies that would serve as hubs for trade, exploration and territorial expansion, including the area now known as New Jersey. Captain Cornelis Jacobsz May, with his extensive knowledge of topography, became the inaugural director of the company. To encourage settlement, he offered free passage and land to individuals willing to inhabit the "New World" for a minimum of six years.

Consequently, settlements were established at Bergen (now Jersey City), strategically positioned along the Hudson River, and New Amsterdam (now New York City), situated at the southern tip of Manhattan Island, during the seventeenth century. These settlements became crucial trade channels for the Dutch West India Company, connecting North America to Europe, and provided a base for Dutch exploration and territorial growth in the area. In addition to these settlements, the Dutch also established a presence in the Lake Hopatcong area, recognizing the area's rich natural resources, despite its challenging climate. The Dutch were particularly enticed by the region's timber resources, understanding their value as trade commodities. These initial settlements set the stage for subsequent European expansion.

The Dutch settlers gained a deep understanding of the challenging, rugged and cold terrain found in the northern region of New Jersey. In contrast, the English initially struggled to grasp the unique environmental characteristics of this land. However, the English displayed early interest in cultivating the area. For instance, in 1634, English explorer Thomas Young erroneously concluded that the region had a Mediterranean climate after discovering grapes.

Conversely, the Dutch possessed a more nuanced comprehension of the region. They accurately recorded the changing weather. They attributed the area's extreme temperatures, distinct from those of Holland, to continental rather than maritime influences, although this understanding might have been somewhat exaggerated. It wasn't until the late 1680s that the English began to develop a more accurate understanding of the region's climate, and by then, the Dutch had already become entrenched in the area.

Johannes De Laet (1581–1649), a prominent Dutch author, played a significant role in acquainting Dutch readers with Northern New Jersey. Johannes De Laet was a noted geographer and a director of the Dutch West India Company. He made significant contributions to the fields of cartography and exploration in his time. One of his notable achievements was documenting the voyage of Henry Hudson to America. His publication, *The History of the New World*, is highly regarded as an exceptional description of the Americas and is considered by some to be one of the finest publications of the seventeenth century. Furthermore, De Laet's cartographic work had a lasting impact, as he was the first to produce maps that included the names Manhattan, New Amsterdam (now known as New York City) and Massachusetts. These maps played a fundamental role in European exploration.

In his written works, De Laet portrayed Northern New Jersey as a promising destination for settlement. His 1621 narrative extolled the region's plentiful wildlife, including deer, turkeys and fish, along with its diverse variety of fruits and inlets. While not entirely precise, his descriptions were remarkable for the era. His literary descriptions of Northern New Jersey laid the foundation for numerous subsequent narratives that continued to pique the interest of readers. These promotional writings were crafted with the intention of encouraging European migration to North America, a prevalent strategy employed by writers during that era. These narratives exemplify a larger pattern observed in colonial literature, providing valuable perspectives on the attitudes of early European settlers.

The Impact of European Settlers on Lenape Society: From Dutch to English

Prior to this era, Native Americans had inhabited New Jersey for tens of thousands of years, fostering unique cultures and communities. Native American communities, like the Lenape, had systems of governance and social structures that allowed them to thrive in their respective regions. They possessed unique languages, art forms and storytelling traditions that conveyed their history, wisdom and worldviews. Various Native American languages existed. The northern New Jersey groups used the Munsee dialect of the Delaware language, while the central and southern New Jersey groups spoke the Unami dialect. The impact of European settlers was nothing short of disastrous as the communities all got lumped together in the eyes of settlers.

Starting with the Dutch, Native Americans longstanding traditions and ways of life were disrupted. The unregulated spread of livestock, particularly hogs, had a profound effect on the ecological balance of the region, leaving lasting scars on the Lenape way of life. These animals roamed freely, causing significant destruction to the forested areas of Northern New Jersey by consuming vast amounts of grass, which was a critical food source for deer and other wild animals. This uncontrolled proliferation of Dutch livestock resulted in the depletion of natural resources and upset the local ecosystem.

The emergence of various diseases, prevalent during European colonization, triggered a catastrophe. The transmission of diseases such as smallpox, measles and influenza from European settlers to Native communities had profound consequences. These illnesses often spread rapidly among Indigenous populations, who had minimal to no immunity against them.

As recorded by Gabriel Thomas, a seventeenth-century observer, the death toll was staggering, with two Native individuals succumbing for every European who arrived in the region. This demographic catastrophe inflicted further devastation on the Lenape community, which was already grappling with the challenges of adapting to the rapid ecological and biological transformations brought about by European colonization. These alterations disrupted the traditional subsistence practices of the Lenape, exacerbating the devastating consequences of the newly introduced diseases and leading to a notable decline in their population.

The deteriorating relationship between the Lenape people and the Dutch compounded the difficulties faced by the Lenape. While the initial contact between the two groups was amicable, it was not long before ministers of the Dutch Reformed Church began to characterize the Lenape people as "savages" or "devils," rather than endeavoring to foster mutual understanding and respect. This attitude among the Dutch clergy undermined efforts to recognize and appreciate Lenape material and spiritual culture, devaluing their way of life and relegating them to a marginal role in Dutch colonial designs. This disregard for Lenape culture and history manifested in a lack of effort to preserve their traditions and promote cultural exchange. Instead, the Dutch dismissed the Lenape as "uncivilized," and in doing so, they perpetuated an unjust power dynamic. Furthermore, the English further complicated the situation, as they sought to challenge Dutch territorial claims. This all added to the Lenape people's instability.

In 1664, the English seized control of Northern New Jersey from the Dutch, which had a significant impact on the region's development. The English brought with them a series of policies that sought to systematically remove the Lenape people from their homeland. Notably, the English settlers displayed a stronger commitment to territorial occupation than their Dutch predecessors, who had primarily focused on trade. Consequently, the influx of English colonizers, pouring in from neighboring regions like Long Island and New England, was much more rapid and aggressive than the earlier Dutch settlement. This resulted in a significant demographic shift that saw the Lenape people gradually become a minority within their ancestral homeland. The Lenape people, therefore, faced an unrelenting struggle to retain control of their communities.

Despite a general disregard for the Lenape, the English managed to overcome the challenges posed by the harsh climate by adopting Lenape practices that were suited to the region. They began to farm in small gardens instead of large fields. This was a safer way to farm, because even if something bad happened to the environment, some of the crops would still grow. This also ensured that they would have seeds for the future. Not all Native practices were adapted. Even though raised fields were common in eastern North America, the English did not usually use them because they were too complicated and labor-intensive. These fields comprised a series of raised mounds separated by canals or furrows, which were thought to help manage drainage and soil fertility. However, other practices, such as raising crops with dirt mounds, persisted. The English also learned from the Lenape to use certain crops to maintain healthy soil. For instance, squash provided a natural mulch, which helped retain moisture and prevent weeds from growing.

Despite benefiting from the expertise of the Lenape people, English colonizers never tried to live in harmony with them, primary focusing on aggressive land acquisition. The English attempted to buy land from the Lenape as a community, using liquor, firearms and various utensils as currency. While the English did not acknowledge Native Americans as landowners, their practices made it more convenient to exploit them.

Early population surveys conducted by the colonizers offer valuable insights into the extent of the Lenape's decline during this period, attributable to disease, English land practices and violence. According to one estimate, there were merely 12 to 30 inhabitants per one hundred square kilometers, or roughly 3 to 8 individuals per ten thousand square miles, culminating in a total population of 2,400 to 6,000 for the state. There is no arguing that these practices disrupted the traditional Lenape way of life, led to the dissolution of their communities and compelled them to relocate to unfamiliar and inhospitable territories.

THE PROVINCE OF NEW JERSEY IN THE LATE SEVENTEENTH AND EIGHTEENTH CENTURIES

In the late seventeenth century, English colonizers established the Province of New Jersey, encompassing present-day Delaware, New Jersey and Pennsylvania. Throughout the colonial period, New Jersey held a limited degree of political influence compared to other English colonies, with its impact on colonial affairs being relatively modest. During this era, a complex web of politics, administrative changes and audacious deception reshaped settlement patterns and land divisions.

New Jersey faced economic difficulties and was distant from the British Crown and Parliament. To address these challenges, the colony began issuing its own paper money, which went unchallenged by the British government and allowed New Jersey's colonial economy to grow. The local government benefited from land interest loans, increased self-governance and appreciated land values. The establishment of the Loan Office Act in 1723 created a government land bank that provided loans to citizens using their land as collateral. However, the colonists' pursuit of land as primary investment capital came at the expense of the Lenape people. Sussex County in Northern New Jersey became a prime location for land speculation in the eighteenth century.

A significant moment occurred on June 8, 1753, when Sussex County was officially established as a separate county, having been carved out of Morris County with the approval of King George II. This act of separation marked a pivotal moment in the region's history. The territory's isolation and abundance of land made it an appealing destination for speculators who were seeking vast tracts of land.

Among these land speculators were notable figures like John and Ann Campbell, who migrated from New York to New Jersey in 1714. They seized the opportunity and purchased 2,761 acres of land in Sussex County. The Campbells were representative of many speculators during this era who aimed to acquire extensive parcels of land. They then sold portions of it and transferred some to their children, thus benefiting from their investment. While land speculators were eagerly acquiring large portions of land, this process had significant consequences for the Lenape people. As more and more land was taken over, the Lenape were pushed out from their ancestral territories at a faster rate.

The lake area was not devoid of individuals of wealth and affluence trying to stake their claim. Among them was Carl Ernst Bertrand, a prosperous German sugar refiner who became captivated by the lake around 1780 due to the iron industry. Bertrand acquired the entire island that now bears his name and constructed a splendid estate on it. This remarkable property featured barns, greenhouses, bowling alleys, stables capable of accommodating twenty horses and a multitude of dog kennels. Before Bertrand's ownership, the island had been in the possession of John Logan of Philadelphia, who served as secretary to William Penn and acquired the island in 1715.

Following Bertrand's death, his wife returned to Germany, and both the estate and the subsequent Bertrand's Island Club met unfortunate fates, succumbing to fires. A club member built a cottage on the island, which was later acquired by Robert Dunlap. In an ambitious endeavor to relocate the cottage to the mainland during a winter day when the lake was frozen, it unfortunately broke through the ice. Nevertheless, the cottage was eventually salvaged, and some of its lumber found purpose in the construction of the old Woodstock House. Mrs. Martha von Campe, one of Bertrand's daughters, became the proprietor of this hotel, and her name lives on in the present-day Villa von Campe located at Chestnut Point.

In 1803, Joseph Bonaparte, the brother of Emperor Napoleon, visited the lake with the intention of finding property to settle in the area. He explored Lake Hopatcong and the Budd Lake region. It is believed that Bonaparte Landing in Byram Cove served as his temporary campsite. Additionally, the Bonaparte family's association with the lake is evident in Elba Point, a site situated on the River Styx, named after the island where Napoleon was exiled.

However, Joseph Bonaparte did not establish permanent residence in the Hopatcong area. According to legend, negotiations for a tract at Budd Lake came to an abrupt end when he chanced upon an unflattering caricature of his renowned brother. Over a century later, the memory of Bonaparte's brief stay at Bonaparte Landing resurfaced when Rex Beach, the esteemed author of adventure stories who resided for many years on the mainland above Raccoon Island, stumbled upon a silver teapot believed to have been left behind by the Bonaparte party.

The Hurd brothers, who established the esteemed Hurdtown Mine prior to 1800, once owned both Raccoon and Halsey Islands. These properties remained within the Hurd family's possession for many decades. Raccoon Island was home to the Hollywood House, a hotel that stood for an extended period before tragically succumbing to a fire in 1912, an unfortunate fate shared by numerous similar establishments.

To add to the unfair treatment of the Lenape, the imposition of English colonizers' landownership practices created misunderstandings, particularly regarding land use and the utilization of resources. At a conference in Easton, Pennsylvania, in 1758, a spokesperson for the Minsi tribe voiced grievances to the governor of New Jersey, alleging unfair treatment by the colonizers, who claimed exclusive rights to wild animals and restricted hunting access. Other tribes also expressed discontent with the Europeans' treatment of the land. The European concept of landownership was foreign to the Native culture, and the complexities of landownership led to misunderstandings and a definitive turning point for the region, forever altering Indigenous culture and traditions.

Disappearance of the Lenape

Through all these struggles for land, the Lenape maintained their resilience the best they could under the circumstances. They sought ways to adapt and survive despite the challenges they faced. But the loss of their land and resources was a severe blow. They made efforts to form alliances with other tribes in Ohio, but the Lenape were rejected. The situation was dire, and the Lenape found themselves in a survival struggle, pitted against other tribes and forced to fight for their existence. One way they fought for survival was through shell bead production, which they used for trade. The Lenape produced millions of shell beads, which became a significant part of their economy. However, competition for resources with other tribes was intense, and they often had to give away their beads to more powerful tribes that threatened them.

By the end of the eighteenth century, the Lenape had virtually vanished from the region. The last agreement between the Lenape and other tribes occurred in 1832, involving the sale of fishing rights, which the Natives had kept out of the sale of Brotherton and other land claims in 1801. The last formal deal involving the entire Delaware tribe and New Jersey took place in 1832. This deal revolved around the trading of fishing rights that the Delaware tribe had retained when they sold Brotherton and other lands to New Jersey back in 1801. The transaction cost $2,000 and was deemed by New Jersey to be a purchase of all land rights from Natives, thereby allowing the state to develop freely. According to the most recent American Community Survey, the Native American population in New Jersey is now only 8,873, which is 0.1 percent of the total population of New Jersey.

Colonial Population Trends

During the colonial period, the state of New Jersey experienced a notably slower pace of development and population growth compared to its neighboring colonies. The state's annual population growth rate averaged a modest 1.46 percent, considerably lower than the usual 4 percent increase observed in other colonial regions. This slower trajectory of growth could be attributed to several factors, one being the prominent influence of English settlers within the state.

English settlers, known for their more conservative colonial policies and practices compared to other European powers, were a significant presence in New Jersey. The governing structures they established often mirrored the hierarchical social and political systems of England, possibly discouraging rapid population expansion. Furthermore, English colonies generally prioritized creating a sustainable agricultural economy over swift territorial expansions, which could have contributed to New Jersey's slower pace of development.

New Jersey's physical environment and geographical location also likely influenced its development rate during the colonial period. The state's topography and soil may not have been as conducive to rapid population growth and economic activity as other regions. These conditions could have posed challenges to the development of large-scale agriculture or other forms of economic production.

Notably, the rate of population growth was particularly slow in the northern region of New Jersey. This trend can be partially attributed to the area's physical geography, including rugged terrain; severe weather; and limited accessibility. However, it's important to note that growth rates varied considerably across the state, influenced by the specific characteristics of each region.

For instance, in the early nineteenth century, the settler population growth in the Lake Hopatcong area, located within the boundaries of Sussex and Morris Counties, was significantly slower than elsewhere in the state. From 1790 to 1810, the population expanded by a mere 0.86 percent. This secluded location, surrounded by hills and forests, may have deterred extensive settlement, resulting in a relatively stagnant population growth.

In contrast, Essex County, located close to New York City and the Hudson River, experienced a more substantial population increase during the same period. The county's settler population grew by 2.49 percent between 1790 and 1810, a rate nearly three times that of Lake Hopatcong. The county's strategic location near a major urban center, access to a vital waterway and its associated trade and employment opportunities likely attracted more settlers, facilitating a more rapid population growth. The population growth in Northern New Jersey during this period demonstrates how different factors, such as geographical characteristics and proximity to trade routes and urban centers, influenced development. The varying growth rates across different regions highlight the intricate relationship between geography and socioeconomic conditions in shaping population trends.

Industrial Population Trends

The period after 1810 marked a shift from the colonial period to the industrial period, and population growth rates also shifted. Sussex County's growth rate dropped to 1.32 percent, while Morris County's grew from 0.86 percent to 2.31 percent, the second-highest population growth rate in New Jersey after Cumberland's 3.16 percent. For further comparison, Essex's growth rate was 1.73 percent. These growth patterns are important to highlight, because they laid the foundation for the development of the Lake Hopatcong area in the subsequent years.

In line with other comparable regions, Sussex and Morris Counties were predominantly male-dominated areas. In 1772, the number of white men per 100 women was 112.78 in Sussex and 117.81 in Morris. This high ratio was due to timber and fishing, which drew men to these regions for economic reasons. Other counties, such as Essex County, had an act to preserve timber, since it was readily abundant in other areas. In contrast to this economically driven influx, Essex County remained largely inhabited by original settlers. As reported by Peter O. Wacker, "In 1750, Daniel Pierson swore that he believed that 'there are scarcely a man in the county of Essex but what is Related by Blood or Marriage to some one or other.'" In comparison to how other counties were developing, the region around Lake Hopatcong was the Wild West—or, in this case, the Wild North—of New Jersey.

Interconnections Between the Past, Present and Future

The contact period also created a nexus between vital components that would shape the lake region. According to legend, a pivotal moment came during the closing years of the seventeenth century, when a Lenape individual nonchalantly showed a piece of rock to a European settler. At that moment, neither the Native person nor the settler could possibly comprehend the profound implications of that simple act. This was a piece of coal. In their daily lives, the Lenape may have used coal, as they were proficient in the use of natural resources. It was from this piece of rock that industry, as well as railroads and the Morris Canal, arose. In terms of iron, the area was an El Dorado. As a result of the coal industry, the Morris Canal was eventually created, paving the way for further industrial advancements in the region.

In parallel, the explosives industry played a significant role in the development of the Lake Hopatcong region, contributing to its transformation and growth. One of the key players in this industry was the Giant Powder Co. of San Francisco, which established its eastern dynamite plant in McCainesville (now Kenvil) in 1871. Afterward, the American Forcite Powder Co. built a plant in Landing in 1883 to manufacture the first forcite (gelatin dynamite) produced in America. The army also set up a powder depot near Kenvil and Landing in 1879, which later became Picatinny Arsenal, and established its powder plant there in 1907.

The presence of the local explosives industry attracted Hudson Maxim (1853–1927), who became a key figure in the Lake Hopatcong region in the twentieth century. Maxim, a distinguished scientist, famously proposed a hypothesis regarding the compound nature of atoms, which aligned with the later accepted atomic theory. In 1888, he became part of his brother Hiram Maxim's gun and ammunition company, where he engaged in experiments involving explosives. Subsequently, Maxim created the Maxim-Schupphaus smokeless powder, a pioneering achievement in the United States, and it gained recognition from the U.S. government. Later, he invented a smokeless cannon powder with cylindrical grains that were perforated, resulting in faster combustion. This groundbreaking invention found widespread application during World War I.

In 1899, the Du Pont industry established an explosive laboratory for Maxim in Landing. Maxim, in his biography, *Reminiscences and Comments*, said, "In 1901, I made arrangements with the Du Ponts, who owned some powder works on the shores of Lake Hopatcong, to come here and erect a laboratory and conduct experiments." As a result of Maxim's collaboration with Du Pont, his lake papers were kept at the Du Pont grounds of the Hagley Center in Delaware rather than in New Jersey. The facility in Kenvil, previously owned by the respected Du Pont Corporation, remained operational as Hercules Powder after Du Pont was divided into three separate companies in 1912.

DAILY NEWS FINAL

25-50 DIE IN N.J. POWDER BLAST

Probe Hints of Sabotage

Above: Maxim testing a weapon. *Courtesy of the Hagley Museum and Library*.

Left: Headline of the explosion at the Hercules Powder Factory. *From "Hercules Factory."*

Opposite: Tribute women sorting out imperfect grains of smokeless powder at Hercules Powder Company's plant in Kenvil, New Jersey, circa 1920. *From Wikimedia Commons.*

On September 12, 1940, a devastating incident occurred within the explosives industry, resulting in a tragic outcome. It was a day that left an indelible mark on the region. The Hercules Powder factory had been operating for many years by then, producing various types of explosives. During World War II, the factory manufactured a mind-boggling 1.8 billion pounds of explosives. However, that day, a catastrophic explosion shook the area and caused devastation.

The most tragic outcome was the loss of life, with fifty-one workers perishing and over two hundred individuals sustaining injuries and burns. The explosions had far-reaching effects, reaching Poughkeepsie, New York, and registering on Fordham University's seismograph, located around fifty miles east of Kenvil. These powerful blasts shattered windows and forcefully tore telephone wires from their poles, even forcing the evacuation of nearby schools. It was so forceful that cars bounced off the roads, windows in homes miles away were shattered and articles flew off shelves and walls.

Dover General Hospital faced overwhelming numbers of victims, necessitating their placement on the hospital's front lawn as they awaited assistance. Injured individuals were transported for treatment in pickup trucks and cars. Doctors, nurses, nursing students and even Boy Scouts from the surrounding area dedicated weeks to caring for burn patients and aiding in locating family members.

The exact cause of the explosion remains uncertain. Speculations include an industrial accident or deliberate sabotage by the Irish Republican Army (IRA) or a group of German Americans in nearby Sussex County. Congressman Martin Dies, the chair of the House Un-American Activities Committee, suggested the involvement of Nazi agents.

Despite the tragedy, the factory was eventually rebuilt, but accidents continued to plague it, and it was finally shut down in 1996. Nonetheless, the sounds of explosions from the nearby Picatinny Arsenal still echo throughout the lake on some days, serving as a reminder of the region's explosive history. Though it is a unique aspect of "lake life," it is often overlooked by those who are used to the sounds and sights of the arsenal. Nevertheless, it remains a significant part of the region's past and present.

Aftermath

The historical narratives of this period are often plagued by fraudulent accounts. These accounts are a result of European biases that have infiltrated into the national consciousness and impacted all regions, and the Lake Hopatcong region is no exception. One such example is the poem "My Mountain Home," which is rife with Euro- and American-centric accounts. The following is the poem as it first appeared in the *Angler*, a Lake Hopatcong newspaper, on June 30, 1894:

> *If we could look back over a stretch of about a hundred years, and let our gaze fall on our beautiful Lake Hopatcong, we should see many strange things, which we can now hardly realize ever happened here, for the shore of Lake Hopatcong were then inhabited by Indians, who were called the Lenno Lenappi. They spent their time in fishing in the clear sparkling waters, and also in hunting the bear, deer, raccoon and other animals. We can hardly think that at one time the shores of our beloved lake were inhabited by these savage people.*

The use of the term *savage* when referring to the Lenape people erases their impressive history and perpetuates the myth of Native incivility.

By examining these narratives critically, we can gain a more nuanced understanding of the social attitudes and cultural biases that shaped them. Another local example of such a narrative is a poem titled "The Indian," published in the *Lake Hopatcong Breeze* on July 7, 1906, which recounts the loss of the Natives:

Lo, the poor Indian, sadly have we seen
His eye dejected and his sombre mien.
The trolley car infests his hunting ground,
His faithful dog is hurried to the pound.
No more around the camp fire does he dwell,
He much prefers to live at the hotel
And take the usual meals of steak and soup
With other members of the Wild West troupe.
Or else behold him in victorious glee—
He does not mourn the hunting ground—not he,
But in illustrious pride steps forth to shine
The favorite player in a baseball nine.

The author of the poem exhibited a deep understanding of Lenape culture and employed this knowledge in creating a prejudiced and explicitly weaponized narrative. Dogs held significant value to the Lenape community and were even interred alongside the deceased. The use of such cultural touchstones to perpetuate harmful stereotypes is particularly insidious. The poem suggests that the Lenape betray their own cultural values by sending their dogs to the pound, a notion that would have been especially significant in a culture where dogs were so highly revered. This type of poetry was highly suggestive and worked to reinforce cultural prejudices and stereotypes of the time at an atomic level. Such narratives can have a lasting impact on perceptions and emotions.

Even Hudson Maxim offered his own biased history, as found in his Delaware archives. From a piece called "Success," written in November 1900:

When the great glacier, deep and vast, lay long ages over all the northern lands, and bleached the smut from the skin of the dark-visaged Indian, making the fair Caucasian, within the flow of great ice-fed torrential floods, by the sun sent to the sea, man fished for food and fought with fierce amphibians; and o'er the frozen desert waste the great cave bear he tracked and killed, or stalked the giant mastodon, or watched for chance to spring with vantage on the sword-toothed tiger, which in turn was hunting him for prey.

> *What a dauntless spirit he must have had—that paleolithic savage, who, armed with but a sliver of flint, entered the dark den and dislodged the fierce cave bear, or slew him there and fed upon the marrow of his bones. Environed by fierce savage life and savage earth and sky, man was cradled in a world of tempest, rocked by forces that placed the spirit and the nerve in the heart of the Viking who ferried the Atlantic in an open boat, and gave invincibility to the Norseman, the giant arms of whose posterity now circle all the earth.*

Maxim took great pride in considering himself a valuable asset to the region and was an actual scientist. However, the evidence presented in this writing suggests that he relied on a fragmented history of the glaciers and Native people to create a simplified and racist decree, which portrayed the development of the area as a seemingly natural progression. Despite Maxim possessing various other skills, this decree was the extent of what he had to offer in terms of the lake's past.

Despite enduring colonialism and its demeaning narratives, Native Americans have managed to resiliently preserve their rich culture and history. I previously mentioned the *manëtu*, but that is just one facet of their multifaceted belief system. Ironically, despite European writers who often labeled them as "savages," there were similarities in the values held by both the colonists and the Lenape. While European Christians sought to convert Natives to their faith, the Lenape had their own deeply ingrained beliefs about creationism.

According to their beliefs, Kishelemukong was the creator god who directed all the life-spirits. As the creator, Kishelemukong sculpted a majestic turtle that eventually grew to become the continent of North America. Kishelemukong also intricately shaped the heavens, the radiant sun, the gentle moon and all living creatures, plants and the cardinal directions that governed the ever-changing seasons. These parallels with biblical accounts challenge the depiction of the Lenape as lacking in spiritual beliefs and further highlight the richness of their cultural heritage, one that could have been appreciated by the Christian colonists.

In essence, these misconceptions about the spiritual nature of the Lenape are representative of the broader issues in understanding and interpreting Native cultures. These distortions, fueled by biased perspectives, have long stood in the way of appreciating the true richness of their societies. The history and culture of the Lenape compel us to confront the urgent need for an honest, unbiased and respectful reexamination of Native histories and their significant contributions. Their story teaches us an important lesson: simply acknowledging the past is not enough. We must actively and responsibly work to correct the distorted narratives and misconceptions that have long overshadowed the representation of these communities and their cultures.

3
THE INDUSTRIAL PERIOD

THE MORRIS CANAL (NINETEENTH CENTURY)

The Morris Canal was a 102-mile-long canal across Northern New Jersey that operated in the early to mid-nineteenth century. It provided an essential transportation link during the Industrial Revolution, carrying coal, iron and other commodities from Pennsylvania to New Jersey port cities. Lake Hopatcong was one of the critical points along the Morris Canal. The lake was a major water reservoir for the canal system, a shipping stop and a recreational destination. The lake provided water to fill the locks that raised and lowered boats along the canal. It marked the industrial phase of the lake.

During the transformative phase when the canal was being built, Lake Hopatcong was reshaped into the vast source of freshwater that still exists today. The Morris Canal became operational in 1831 and was in use until 1924, resulting in the lake's present-day size and shape. The economic downturn in Northern New Jersey, combined with its abundance of natural resources, made the construction of the Morris Canal a sensible option. Because canals provided an efficient means of transporting goods and people, they played a crucial role in the eighteenth and nineteenth centuries. They were among the fastest and most reliable ways to move large quantities of goods over long distances before modern railways and roads.

The Morris Canal was a massive engineering feat that consumed six counties: Hudson, Essex, Passaic, Morris, Sussex and Warren Counties. The canal ran from the waters of the Delaware River, near Easton, Pennsylvania, to the Hudson River Harbor in New Jersey, passing through areas like Jersey

A canalboat, New Jersey. *Courtesy of the Hagley Museum and Library.*

City. However, the varying elevations of these areas presented significant challenges to the canal's construction and operation. As Maxim once observed, the Morris Canal was "a mountain climber. "

The highest elevation of the canal was around Lake Hopatcong, which played a crucial role in making the development of the Morris Canal feasible. It was not only the most critical resource for the canal, supplying most of its water, but it was also one of the toughest territories to traverse. The Morris Canal rose 914 feet near Lake Hopatcong and then descended 760 feet to its lowest level at Phillipsburg on the Delaware River. Despite the challenges, the Morris Canal was a triumph of engineering that transformed transportation in New Jersey.

From Concept to Waterway

Before the Morris Canal was ever heralded as an innovation of industry, it was first a fishing spot for one man. George Perrott MacCulloch was the president of the Morris County Agricultural Society. As a young man, MacCulloch immigrated to the United States from Scotland. In June 1822, George MacCulloch, then a resident of Morristown, New Jersey, purportedly conceived of the Morris Canal while fishing at Lake Hopatcong, located in the north-central highlands of New Jersey.

Much like the iconic moment of realization experienced by Newton when an apple fell from a tree as he sat beneath it, MacCulloch's inspiration also struck suddenly. While fishing, he peacefully observed the serene flow of water from the Delaware River merging into the Passaic River. In that tranquil moment, a visionary idea took root in his mind—the idea of connecting these waterways to create a navigable route to New York City. This vision held the promise of rejuvenating the ailing iron industry in Morris County, offering a lifeline to the struggling local economy and transforming the region's transportation landscape.

However, in Barbara Kalata's publication, *A Hundred Years, a Hundred Miles: New Jersey's Morris Canal*, she raised doubts about the accuracy of MacCulloch's anecdote. Kalata contends that discussions about the canal had already taken place, even dating back to William Penn. But no matter how MacCulloch came up with the idea, he started planning the Morris Canal, thinking that Lake Hopatcong would make a good water supply. He considered routing the canal to follow the easiest and most economical path. Despite uncertainties regarding the Sussex County terrain, he believed the project was feasible.

Regardless of the origin of the idea, MacCulloch assumed a leadership role in the Morris Canal campaign. He asserted that the Great Pond, now known as Lake Hopatcong, would serve as an ideal and abundant water reservoir for the canal's highest point, situated between Rockaway and Muskonnetconk. MacCulloch initiated his public campaign during the summer of 1820, intensifying his efforts in 1821 by publishing numerous pro-canal articles under his own name in New Jersey newspapers.

MacCulloch staunchly refuted any detractors who argued that the canal's advantages would be limited to just one part of New Jersey, dismissing such thinking as "selfish." He ardently advocated that the canal's benefits would extend to benefit the entire state, highlighting his concerns about the lack of unity among its residents. MacCulloch eagerly shared his grand vision with the public, urging them to examine their maps and appreciate the widespread advantages of the project. Furthermore, he emphasized how the Morris Canal would hold a distinct advantage over the Erie Canal, with a boating season that would extend for an additional six weeks.

In a bid to bolster the appearance of widespread support, MacCulloch employed pseudonyms, akin to contemporary anonymous social media accounts like Twitter/X burners. While the primary purpose of the Morris Canal was to facilitate the transportation of coal and iron, MacCulloch utilized these pen names, "Agrestis" and "Aristides," to advocate that the canal would also yield substantial benefits for other sectors, such as apple farmers. He argued that this development would enable these farmers to transport their cider to New York City, humorously cautioning against an excess of whiskey that they couldn't possibly consume alone. MacCulloch's campaign proved so effective that public sentiment overwhelmingly favored the canal's construction. He even ventured to predict that people might be so enthusiastic about the canal that they would willingly surrender their land for the project. However, this last prediction didn't quite materialize as anticipated.

The journey toward realizing this ambitious project was riddled with hurdles, most notably the challenge of navigating the previously mentioned treacherous terrain of Sussex County with the technological limitations of the time. Nevertheless, MacCulloch had assembled a group of supporters for the Morris Canal who remained steadfast in their commitment to persuading the state legislature that this transformative undertaking was worth pursuing, even in the rugged mountainous regions of Northern New Jersey. Unfortunately, the champions of the canal also encountered the challenge of "sectionalism," as southern counties initially resisted collaboration with their northern counterparts on such a monumental project.

MacCulloch's tireless endeavors culminated in a significant milestone on November 14, 1822, when the state legislature passed a bill with an overwhelming majority of 27–9 in favor of constructing the Morris Canal. This resounding endorsement from legislators clearly underscored the substantial support that had been garnered for the canal's development. In 1824, following a comprehensive assessment of the Lake Hopatcong area, private investors established the Morris Canal and Banking Company to provide the necessary funding for the project. As a result, a select group of influential individuals from Morris County was appointed by the legislative body to thoroughly examine the potential canal route and assess its viability.

Several potential routes were proposed for the eastern section of the Morris Canal, each differing in terms of cost, path and termination points. Evaluating these options also involved an examination of their water supply requirements. Engineers Benjamin Wright and Canvass White were initially tasked with the selection of the most favorable route but faced difficulties in reaching a decision. In 1825, Ephraim Beach assumed the role of chief engineer and undertook the responsibility. Beach assessed the available options and presented a comprehensive report outlining four potential routes: a southern route passing through Morristown and Elizabeth, two middle routes via Troy and either Boonton and Pine Brook or Newark and a northern route via Boonton and Pompton Plains.

Notably, Beach's approach in this evaluation was pragmatic, as he refrained from expressing personal opinions and instead focused on practical considerations. After careful assessment, he recommended the northern route, ending at the Passaic River, primarily due to the area's burgeoning industrial development and its promising transportation potential. This included the prospect of connecting to the Pompton River. He believed that this route would best serve the interests of both the public and the company, particularly benefiting the iron industry.

Following his study, Beach arrived at a crucial conclusion: the feasibility of the project was intricately tied to the existence of Lake Hopatcong, strategically located at the summit. This lake held the potential to furnish the canal with a generous supply of water in both directions. In fact, his study indicated that Lake Hopatcong could offer three times the amount of water required by the canal. These findings substantiated MacCulloch's claims that the "Great Pond" was, without a doubt, the linchpin that made the Morris Canal project not only feasible but also highly workable.

Another critical factor that contributed to the possibility of building a canal during this time was the alleviation of the labor shortage that had plagued early nineteenth-century America. This scarcity of labor resources was a multifaceted issue influenced by various factors, and its ramifications were profound for the development of new projects. As the United States experienced rapid expansion, the demand for labor consistently surpassed the available workforce, particularly in growing urban and industrial areas. The westward expansion of the nation attracted a growing number of individuals to new frontiers, where they played pivotal roles in cultivating uncharted lands.

However, the influx of Irish immigrants, who had been arriving in substantial numbers since 1819, emerged as a critical source of labor. In addition to constituting a substantial workforce, Irish immigrants were known for their strong work ethic and willingness to put in long hours. They played an essential part in constructing much of America's infrastructure during this period, making them vital contributors not just to Lake Hopatcong, but to the growth and development of the nation.

With the project underway, the Morris Canal and Banking Company (MC&BCo.) made its shares available to investors and speculators in the spring of 1825. On March 16, a flurry of activity took place at Hayden's Tavern in Morristown, where many prospective buyers lined up for hours to purchase the shares. Eager investors submitted subscriptions for banking

Man-made water barriers. *Author's collection.*

shares, and the demand surpassed the available stock. The company's charter granted authorization for issuing one million shares, each priced at seven dollars per share.

The initial sale of stocks, despite the hysteria and confusion it generated, successfully raised the funds needed for the initial construction. The men working on the canal were compensated with company bills. By October 5, around seven hundred men were employed at the canal, and excavation work at The Great Pond (Lake Hopatcong) was mostly completed during the winter.

The construction of the Morris Canal did not adhere to the customary tradition of commencing on July 4, which was considered an ideal time for such projects, particularly canal construction, in the United States. Instead, the official beginning of Morris Canal construction occurred in mid-July 1825. The canal's dedication ceremonies took place on October 15, 1825, with around eight hundred invited guests in attendance. The groundbreaking ceremony transpired at Lake Hopatcong, drawing people from New York, Pennsylvania and New Jersey to hear William Bayard, the president of the Morris Canal and Banking Company, deliver a speech on behalf of the directors.

Even with the construction underway, resistance to the canal project persisted. Engineer John Langdon Sullivan, situated in the more populated regions near the Passaic River, notably opposed the canal project expanding into his area. Sullivan, famous for inventing the steam towboat and advocating for the Paterson region, in *Refutation of Mr. Colden's "Answer" to Mr. Sullivan's Report to the Society for Establishing Useful Manufactories in New-Jersey upon the Intended Encroachments of the Morris Canal Company in Diverting from Their Natural Course the Waters of the Passaic*, published in 1828, expressed surprise at the lack of opposition from the city of Paterson toward the canal's expansion, so he championed the cause.

Sullivan raised concerns about the constitutional legitimacy of the Morris Canal's control over the waters of the Passaic River. His defense primarily focused on local manufacturers and the industries reliant on the river's water, rather than advocating for a specific environmental standpoint. Sullivan's argument centered on the idea that the canal should be located only in the rural areas around Lake Hopatcong, highlighting the lake's ample water supply, eliminating the necessity of diverting water from the Passaic River. Sullivan declared that the canal company had intentionally planned to redirect Passaic River's waters away from the manufacturers in Paterson and had already initiated construction to achieve this objective. He cited how the waters of Lake Hopatcong had been demonstrated to be more than sufficient to support the canal. Nevertheless, in the event that they proved insufficient, Sullivan proposed establishing a rail line connecting Philadelphia to the Hudson River as a viable alternative, eliminating any necessity for the Passaic River.

However, the Morris Canal Company ultimately prevailed in using the Passaic River and continued to expand its operations. The Passaic River played a significant role in the Morris Canal system. Designed to link the Delaware River in Phillipsburg, New Jersey, with the Hudson River in Jersey City, the Morris Canal followed a carefully planned route that incorporated several waterways. As an integral part of the overall canal route, the Passaic River facilitated the transportation and navigation of goods and boats, enabling the successful and expansive operation of the Morris Canal.

CONSTRUCTING THE CANAL

Constructing infrastructure at high elevations required the use of different methods. A comparison between the construction of the Erie Canal, finished in 1825, and the construction of the Morris Canal highlights the significant challenge faced by the builders of the Morris Canal. The Erie Canal, which stretches across the state of New York, had a maximum elevation of around 568 feet above sea level near Rome, New York. The Morris Canal had to climb an impressive 914 feet between Newark Bay and Lake Hopatcong, and then it had to descend 760 feet to its end at Phillipsburg. Overall, the total change in elevation along the Morris Canal was a substantial 1,674 feet. If the Morris Canal had used the same lock ratio as the Erie Canal, it would have inevitably slowed down water transportation.

The process of coming up with strategies to overcome the elevated terrain involved a series of well-planned and timely design improvements. For example, MacCulloch initially suggested using over two hundred traditional lift locks, each with an eight-foot change in elevation, to help navigate the ascent into Lake Hopatcong. However, James Renwick, a renowned expert in natural and experimental philosophy at Columbia University, proposed that inclined planes were a more suitable and realistic solution. Renwick's designs were acknowledged and awarded a silver medal by the Franklin Institute of Philadelphia in 1826. Considering the geographical features of the area, inclined planes emerged as a practical and workable choice.

In 1827, the idea of inclined planes was officially endorsed. Cadwallader D. Colden, the president of the Morris Canal and Banking Company, followed the recommendations of engineers from the United States government, who had surveyed the canal route, as well as other experts consulted by the company. He wrote in *A Report to the Directors of the Morris Canal and Banking Company* that "the Engineers of the United States, who surveyed our canal route, under the directions of the General Government, as well as all the persons the Company has employed or consulted, have given decided opinions in favor of inclined planes." Despite this initial endorsement, the first attempts to implement them were not successful. Nevertheless, Colden maintained his belief in their potential, asserting that there was "no reason why [there] should not be entire confidence in the planes." Engineer David Bates Douglass from West Point later played a pivotal role in vindicating Colden's conviction. Between 1829 and 1832, Douglass reengineered the inclined planes, rendering them functional and efficient.

It was through such innovation that the construction process of the canal garnered admiration from engineers and scientists alike. *Scientific American* published a report on the canal's pioneering design and how its construction marked the first instance of the inclined plane's usage within the United States. The report boasted about how these planes facilitated the transportation of boats over high elevations. Operated by waterwheels, boats were divided into two separate compartments, connected by latches and pins, with each compartment functioning independently. The boats, typically weighing around sixty-five tons, could then be transported by divided, eight-wheeled trucks across inclined tracks. This innovative design and infrastructure allowed for efficient transportation along the Morris Canal inclined planes.

Despite the greater cost efficiency of locks, the inclined planes offered a solution to the challenges of time and efficiency when crossing the canal. The entire journey could be completed in approximately five days, and minimal manpower was required, as each plane system was overseen by an individual responsible for applying the brakes.

The efficacy of the inclined planes stemmed from their utilization of hydraulic locks, which facilitated the ascent of boats and their cargo on an inclined car. Segner turbines, also known as reaction or Scotch turbines, were constructed to drive the inclined plane lifts on the Morris Canal. A four-and-a-half-foot-tall water wheel, powering an eight-inch rope, swiftly pulled the boats up the incline, enabling a remarkably efficient transfer time of merely four minutes. The canal incorporated a total of twenty-three inclines equipped with hydraulic lifts, covering a substantial portion of the canal's route. Notably, over five miles of the canal were constructed using rail, demonstrating a well-organized and streamlined engineering accomplishment. Consequently, Lake

A cable joint located at Inclined Plane 9 West in Phillipsburg, New Jersey. *From Wikimedia Commons.*

A canal lock and basin in Boonton, New Jersey. *Courtesy of the Hagley Museum and Library.*

A Scotch turbine from the Morris Canal. *Author's collection.*

Hopatcong underwent a transformative shift in its appearance. No longer solely characterized by its picturesque hills and woodlands, the lake acquired an infused industrial aesthetic.

In the 1850s, the Morris and Essex Railroad extended its operations to Lake Hopatcong but did not build a station there. Passengers had to disembark at the Drakesville station and endure a stagecoach ride to the lake. In 1866, the Ogden Mine Railroad transported iron and zinc ore from Ogdensburg to a transfer terminal at Nolan's Point. The ore was then loaded onto canal boats and towed by steam tugs to the Morris Canal feeder, where mules pulled the boats into the main canal for delivery to foundries. The railroad network eventually bypassed the canal, but in 1882, the Central Railroad established a passenger station at Nolan's Point, attracting tourists with excursion trains. This led to the development of amusements and a thriving tourism industry, offsetting the declining mining business.

During the 1870s, MacCulloch astutely recognized the burgeoning tourism industry's potential and seized the opportunity by founding the Hopatcong Hotel. In a remarkably short span, he transformed it into a highly coveted destination for travelers. Nestled on the lake's western shore, the hotel is said to have provided guests with spectacular views of both the serene lake and the imposing greenery in the background. The establishment was also known for its luxurious features. This included gourmet dining, an opulent ballroom and a spacious veranda designed for leisure and relaxation. Additionally, an array of recreational activities, including boating, fishing and swimming, were readily accessible, further enhancing the hotel's appeal as a retreat for visitors hailing from New York City and the neighboring regions.

The Morris Canal's Pricing System and Protective Measures

The Morris Canal faced substantial expenses during its construction. The total cost of building the canal, including all the work completed by 1831, amounted to an astonishing sum exceeding $2 million. To offset these significant expenditures, the Morris Canal and Banking Company devised a well-structured pricing system. The primary source of revenue for the canal derived from the transportation of coal and iron. In 1866, an impressive 889,220 tons of cargo were transported through the canal, with coal accounting for half of the total volume.

Other commodities transported included iron, zinc and a diverse range of goods, such as cider, vinegar, beer, whiskey, wood, sugar, lime, bricks, hay, hides, iron ore, lumber and manure, albeit on a smaller scale. As envisioned by MacCulloch, the canal even served as a beneficial means for merchants to transport their inventory. The canal's tariff structure accommodated an array of smaller cargo that could be exported, encompassing items such as bran, butter, bricks, coffee, corn, cotton, ashes and pots, among numerous others. Additionally, the canal offered transportation services for passengers at a rate of half a cent per mile. The pricing system was aimed at being clear and keeping the canal financially viable.

However, protecting the substantial investment against potential damage was also a crucial concern for the company. In 1835, they released the "Rules and Regulations of the Morris Canal," which outlined stringent measures to safeguard the canal's infrastructure. One particular measure focused on preventing boat damage to the locks and gates. According to the regulations, every boat passing through a lock was obligated to secure a bow and stern line in a way that prevented any contact with the gates. The company emphasized the importance of this precautionary measure to ensure the integrity of the canal's operation. Neglecting this requirement and causing damage to the gates or locks resulted in the boat's owner becoming liable for the resulting damages, in addition to facing legal penalties.

The implementation of this extensive policy showcased the company's dedication to safeguarding the infrastructure of the canal and ensuring the protection of its significant investment. Through the enforcement of rigorous measures and holding boat operators responsible, the Morris Canal and Banking Company sought to uphold the canal system's long-term functionality and sustainability.

Rising from the Depths: The Metamorphosis of Lake Hopatcong

The Great and Little Ponds, remnants of the ice ages, constituted notable geographical features in the region. The Great Pond encompassed a significant portion of the area, while the Little Pond represented a comparatively smaller body of water located to the north. The two ponds were separated by two miles. Over time, these ponds, particularly the Great Pond, underwent distinct modifications. Around 1764, the construction of a dam at the southern end of the Great Pond, designed to support the Brookland forge and mill, resulted in a rise in the water level by approximately six feet. This alteration potentially led to the formation of islands after some of the original lake's protruding points were submerged. This adjustment facilitated the subsequent joining of the lakes by ensuring a higher water level.

In 1825, the Morris Canal made its initial impact on the region, as contracts were awarded for the construction of the western segment of the canal, spanning from a location approximately seven miles east of the Delaware River to the peak level at Lake Hopatcong. From there, the canal extended eastward to the town of Rockaway, where the construction of the first experimental plane was underway. These significant developments solidified the foundation for future transformations, ultimately resulting in the inclusion of the Great Pond and Little Pond within the scope of the Morris Canal project.

The Great Pond underwent several stages of elevation, gradually evolving into the present-day lake. Significant transformations occurred at the northern end of the lake basin, leading to notable changes in the landscape. The land between the Great and Little Ponds became submerged, causing Raccoon and Halsey Islands to be cut off from the mainland. Along a narrow depression, the water flowed farther, giving rise to Prospect Point and effectively isolating it from the land north of Nolan's Point. As the water level rose, it created a clear division between Liffy Island and the mainland, incorporating Little Pond into the larger body of water that we see today.

Moving toward the southern half of the lake, the inundation flowed back around Pickerel Point, forming the distinctive River Styx. Closer to the dam, a new feature emerged known as King Cove, while the previously named Canfield Island, now Bertrand Island, became separated from the mainland. Another low-lying area to the south of the dam was also affected by flooding, resulting in the formation of a narrow and shallow cove that stretches toward Landing.

Lake Hopatcong actively resisted human endeavors throughout the process. A winter storm in 1827 damaged one of the locks, posing a threat to the entire operation. However, the workers remained undeterred, swiftly repairing the damage, showcasing their resolve against nature. Mastering the lake required continual attention and maintenance—an ongoing battle. By April 1845, approximately 1,200 workers were laboring tirelessly, racing against time to complete the project by June 1. The creation of Lake Hopatcong was no easy feat, but the ingenuity and determination of the workers made it possible.

The impact of the canal extended beyond its construction phase, altering the fabric of the surrounding region after its completion in 1832. The transformation led to the replacement of woodlands with barren lands. The lake itself exhibited symptoms of change that affected both the people and the land, including increased rain runoff and evaporation, resulting in the drying out of numerous streams and brooks.

Hazards of the Canal

During the nineteenth century, building canals was a difficult and labor-intensive process. Technological advancements were limited. Wheelbarrows were the largest vehicles used, while picks and shovels were the most common tools. Clearing trees and moving large rocks were highly labor-intensive tasks of that era. Horse-drawn scrapers were occasionally employed, but the scarcity of draft animals limited their effectiveness. As farmers observed the mistreatment and unfortunate fate of their beloved horses, their willingness to provide their livestock for the construction of the canal gradually diminished. Initially driven by the promise of financial benefits, they began to prioritize the emotional toll and the detrimental effects the work had on their livelihoods. The negative impact outweighed any incentives that were offered.

In 1854, the escalation of water levels posed a significant hazard in the vicinity of Lake Hopatcong. Residents near and around the lake experienced substantial property damage due to the ensuing flooding. Faced with this pressing issue, concerned residents united to formally file a complaint, urging the Morris Canal authorities to take immediate action. An article published in the *New York Daily Times* extensively detailed the extent of the damage suffered by nearby roads and lands as a consequence of the rising waters. The company's officers claimed that they were unaware of these issues, and while they helped fix property, not much was done to prevent this from happening again.

In 1889, heavy rainfall led to higher water levels again, causing further outcries by residents. Consequently, experts in dam construction and water management were summoned to assess the situation. The experts meticulously assessed the dam to determine whether it could handle the mounting water pressure. The findings of the expert assessment validated the concerns of the local residents. It was evident that the existing dam, faced with the surging waters from heavy rainfall, posed a significant risk to both the lake's stability

Current Lake Hopatcong water flow. *Author's collection.*

and the safety of the nearby communities. In response to these alarming revelations, a comprehensive plan was developed for the construction of a new dam. This new dam served two purposes: first, it alleviated the mounting water pressure within Lake Hopatcong and second, it protected the surrounding communities from potential flooding and property damage. The construction of the new dam not only provided relief from immediate threats but also laid the foundation for the continued development and resilience of the region in the face of future environmental challenges.

Echoes of the Past: The Enduring Legacy of the Mighty Canal

In her novel *A Summer's Adventure on the Morris Canal (Early 1900s)*, Mary Thompson offers a unique perspective on the canal fifty years after its closure. Rather than solely focus on the canal's economic and engineering aspects, she delves into the local perspective of the people living near the canal. Through the character of Joey, a young man who leaves behind his past generations to explore the new world of industry, the canal becomes a frontier. Thompson portrays the canal as an integral character in the region, harmoniously blending with the area's natural bodies of water. She anthropomorphizes the canal, framing it as a victim of time, and concludes that its spirit still lives with us.

Thompson's novel presents a male perspective, with the character of Effie briefly inserted into the story as a reminder of women's perspective. Effie, a red-headed girl who wants to play, is disregarded by Joey and viewed as a nuisance by his grandfather, who reminds Joey of his more serious business.

Thompson paints a vivid picture of the Morris Canal's history as a story of human perseverance and dedication. Instead of focusing on the canal's engineering feats, the captain in the story marvels at how the canal was dug entirely by hand. Thompson masterfully uses this detail to evoke a sense of nostalgia and romanticism for her readers. She highlights the sheer human effort that went into creating this marvel of transportation.

Thompson's approach to storytelling is a powerful reminder of the influence that stories can have on a community's consciousness. By framing the canal's history in terms of human sweat and hard work, she effectively captures the imagination of her readers and draws them into a world that is both gritty and inspiring. Her narrative offers a window into a bygone era, allowing us to appreciate the struggles and triumphs of the canal.

4

THE BOOM

UNLEASHING A NEW LAKE IN THE LATE NINETEENTH CENTURY TO THE EARLY TWENTIETH CENTURY

In this chapter, the exploration into the past has been made possible through the invaluable resources provided by the Hagley Museum and Library in Wilmington, Delaware. The archival references, including box, folder and accession numbers, refer to the collection of papers available, particularly those belonging to Hudson Maxim. For this book, these historical documents have offered a window into the past, enabling a deeper understanding of Maxim's pivotal role in shaping the lake and many of the issues relevant to the region at the time. By delving into these papers, researchers and scholars gain access to a wealth of information, allowing for a comprehensive analysis of many different subjects, which are particularly described on the Hagley Museum and Library's website. The generosity of the Hagley Museum and Library in making these resources available is admirable. It provides a valuable opportunity for individuals from various disciplines, including historians, economists, sociologists and environmentalists, to delve into primary source materials.

LAKE HOPATCONG: A NEW HAVEN

During this period, Lake Hopatcong transformed into a hub for recreational pursuits, including boating, angling and swimming. As far back as 1895, the Delaware, Lackawanna and Western Railroad Company notably lauded Lake Hopatcong, contending that it bore such resemblance to Lake

George that it was frequently referred to as the "Lake George of New Jersey." Leveraging the widespread allure of Lake George, the railroad company was intent on casting a similar light on Lake Hopatcong. The company persisted in this comparison, acknowledging the considerable size disparity in its *Summer Excursion Routes and Rates* release. It read, "This allusion is justified, although Lake George is so much larger as to make the comparison hardly fair." Despite this caveat, the publication unequivocally concluded that Lake Hopatcong's scenic beauty was unparalleled. The narrative painted an idyllic atmosphere, ideal for camping and leisurely activities, such as reading and singing songs around a campfire. It was promoted as an ideal location for quality time with family and friends, offering a perfect respite from the incessant demands of everyday life. This was clearly an attempt to market the area as a desirable destination for relaxation and enjoyment.

From this point, Lake Hopatcong's history acquired a certain air of romanticism. A vivid illustration of this can be found in the 1919 pamphlet published by the Lake Hopatcong Chamber of Commerce titled *Beautiful Lake Hopatcong: What It Is, Where It Is and How to Get There*. The pamphlet,

A backyard campfire at Lake Hopatcong homes. *Author's collection.*

found in the Hagley Center archives, offered a positive and evocative depiction of the lake, intending to attract potential visitors:

> *Deep in the wooded hills of Northern New Jersey, a great, white jewel set in green, Lake Hopatcong calls with cool sweet promise to the vast army of men and women retreating under the blazing guns of General Summer. Lake Hopatcong! It is an indescribably beautiful spot, unspoiled by too much "improvement" yet providing every comfort and modern convenience for visitors; in places as wild and primal as it was in the days of the Lenni Lenape, yet in immediate touch with the great arteries of business so near the world's commercial heart that the contrast excites new wonder.*

The pamphlet aimed to market the area by illustrating a symbiotic relationship between people and the natural environment. A significant figure in shaping this era was Hudson Maxim.

Hudson Maxim

Hudson Maxim was an esteemed American inventor and chemist who was born on February 3, 1853, in Orneville, Maine. His notable achievements lie in the realm of explosives and munitions. Beyond this field, he made significant advancements in diverse areas, such as aviation, gas engines and fire suppression systems. His innovative prowess resulted in numerous patents and widespread recognition. Hudson Maxim's remarkable contributions to science and technology have left an enduring impact across multiple industries.

On August 14, 1896, a year after the railroad company trumpeted the virtues of Lake Hopatcong, Hudson Maxim, the inventor of the first smokeless gunpowder, couldn't wait to brag to his mother about his new wife, Lillian. (This information is from Hagley, accession 2147, box 1, folder 13.) She was his second wife, as he had, at some point, divorced his first wife—legally, hopefully. Not one to be indecisive—or patient, for that matter—Maxim was engaged in only three weeks, so his family in America knew nothing of this news. Writing from his travels in England, he apologized to his mother for not letting her know sooner, but he bragged that from Lillian's photograph, his mother would "understand how it happened." It was not her eyes or any other feature that he wanted his mother to see. After all, "[e]ven a dog has pretty eyes." Lillian had a pretty mouth, a sophisticated feature on his "obedient" wife. "The mouth being the last thing that received a high development in man, the degree of development of the mouth tells the character of the person more than any other feature." Maxim could rest assured about his choice in companionship, because Lillian had "the prettiest mouth that I ever saw."

After extolling the qualities of his new bride, Maxim took note of Lillian's father, a clergyman. It was not her father's faith or preaching that Maxim described but how he wrote for newspapers. Maxim bragged, "He is a great teetotaller, but none of the family can resist a Mint Julep. He is a great linguist; he reads and speaks twelve different languages." Maxim was drawn not just to the appearance of Lillian's mouth but to what came out of it. His future journey would bring him from England to Lake Hopatcong, where his mouth and linguistic skills would play a significant role in the lake's history.

Simultaneously, Lake Hopatcong, hitherto unknown to Maxim, was beginning to establish itself as a coveted summer retreat. Evidence of this burgeoning status can be found in the August 15, 1896 issue of the *Angler*. The Lake Pavilion Hotel, one of four such establishments in the area at the time, offered daily dancing, ice cream and an abundance of soda for $2 a day or $12 a week. Furthermore, it hosted Challenge Cup Races from its dock. For those disinclined to long-term stays, cottages were available for purchase at $5,300. The amenities included daily steamboat cruises, rental canoes and bathhouses and catering to both transient and permanent visitors. This trend significantly diverged from the historical use of the lake by campers who primarily visited for hunting and fishing. In fact, the number of camps had been significantly reduced, from a historical high of around one hundred to just about twenty at the time.

The expansion of train service into the Lake Hopatcong region played a pivotal role in accessibility and its development as a popular summer

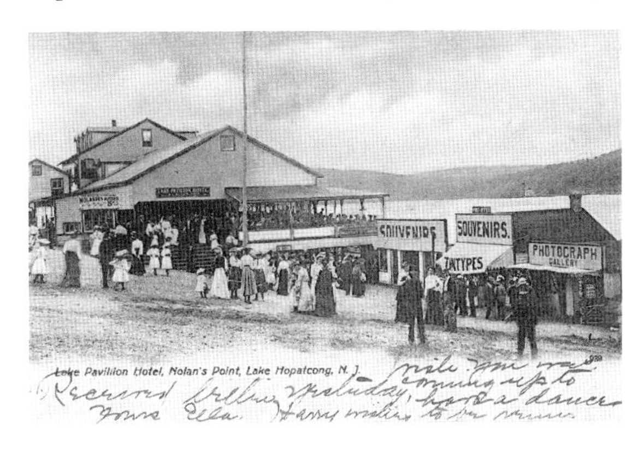

The Lake Pavilion Hotel, circa 1900. *From Wikimedia Commons.*

destination. As cities experienced rapid industrial expansion during that time, their environments suffered from polluted water supplies, contaminated air and overcrowded living conditions for the working class and their families. Seeking a refuge from these environmental degradations, individuals turned to areas such as Lake Hopatcong.

By the end of the nineteenth century, Lake Hopatcong was primarily a resort area. Motorized vessels replaced traditional rowboats and canoes, introducing a new form of pollution to the lake. The number of accommodation facilities increased from four hotels in 1896 to "more than forty hotels and boardinghouses as well as several hundred summer cottages" along the lake's fifty-mile shoreline. This shift from camping and canoeing to more leisurely pursuits marked the lake's transition to a recreational hub.

The character of the lake catered to individuals of Hudson Maxim's social standing, not just modest visitors or fishermen. The shoreline witnessed rapid growth in the construction of cottages and businesses, resulting in a progressive limitation of public access to the lake. The proliferation of summer resorts and businesses reduced the availability of open swimming and fishing areas. This development trajectory caused a shift in the lake's ambience, transitioning from a local amenity to a more exclusive destination.

In the spring of 1886, Hotel Breslin, which became the Alamac around 1918, opened its doors above the crossroads of Edgemere and Windermere. Standing majestically tall until it was destroyed in a fire on February 21, 1948, it was a lighthouse of luxury. It was a symphony of elegance designed by the legendary Frank Furness. Its verandas, gardens and lake views were a visual feast for its guests. This four-story, one-hundred-room luxury palace also had the distinction of being the first electrified building on the lake. In 1896, it transformed into the private Lake Hopatcong Club, an experiment that ended in financial difficulties, leading to its reversion to the Hotel Breslin after three seasons. Breslin wasn't just a footnote in Lake Hopatcong's history—it was a milestone. With its establishment, the lake burst onto the radar of celebrities and the wealthy, further ushering in Lake Hopatcong's golden era of glamour.

The Breslin's popularity spurred the development of several other notable hotels. At the dawn of the twentieth century, over forty hospitality establishments and boarding facilities, along with several hundred seasonal residences, proliferated along the fifty-mile periphery of the lake. Among them, Justamere Lodge, established in the early 1900s by Henry W. Tietjen, stood on a hill overlooking the lake in the town of Landing. The lodge offered numerous amenities, including a ballroom, a dining room and a swimming pool, making it a popular vacation spot for several years. However, as the lake's popularity waned in the mid-twentieth century, the lodge closed and was repurposed. Despite no longer being operational, Justamere Lodge remains an integral part of Lake Hopatcong's history and development.

Another notable establishment was Hotel Westmoreland in Mount Arlington, designed in the Victorian style and offering a myriad of amenities. During its prime, it attracted numerous celebrities and dignitaries. Other hotels, such as Hotel Arlington in Mount Arlington and Mohawk Hotel in Sparta, offered modest accommodations that appealed to families and budget travelers. These hotels and resorts were a significant part of the local economy and community and played an essential role in Lake Hopatcong's evolution in the 1920s.

During the Prohibition era, Lake Hopatcong was not devoid of vices. Despite the nationwide ban on alcohol, individuals visiting the lake, including theatrical performers and others, did not have to endure thirst. Speakeasies operated discreetly all around the lake, ensuring that people could partake of illicit alcoholic beverages. One speakeasy, located near the River Styx, even offered a convenient curb service for boaters, swiftly preparing a quart of gin while the customer waited.

The Stars of Lake Hopatcong

The lake began to attract a diverse array of luminaries. The early twentieth century was termed "the Joe Cook era" by Jill P. Capuzzo in her *New York Times* article "We Call It Lake Life." Cook, a charismatic Broadway celebrity known for hosting grand parties, was a regular visitor, with notable guests such as Babe Ruth (evidence of whose visit remains inscribed on Cook's piano, currently housed in the Lake Hopatcong Museum). As Capuzzo notes:

> *The Joe Cook era marked Hopatcong's glory days, when vacationers in the early 20th century stayed in grand hotels, boated on the lake and visited Bertrand Island Amusement Park. After plane travel opened up more exotic locales, the area fell on hard times, and many of the old homes and hotels burned down or were otherwise destroyed. But the lake is now enjoying a second act, as more people have discovered its vacation-like charms during the pandemic.*

This was a period that shimmered with the luminescence of stars, and it is the one most often discussed when talking about Lake Hopatcong. It was a time of dazzling personalities—the zenith of star power—and perhaps Cook was the brightest.

Emerging from Evansville, Indiana, in 1890, Joe Cook was on a path to stardom. A maestro of multifarious talents, Cook's artistry ranged from tightrope walking to miming. He spun outrageous stories that left audiences in stitches, while his dexterity with various musical instruments amplified his entertainment prowess. Known as the "One Man Vaudeville," his versatility was legendary. Cook's magic wasn't confined to vaudeville.

Joe Cook gained significant fame through his roles on Broadway, notably his leading role in *Rain or Shine*, which ran from 1928 to 1929. His performance was so well received that he reprised his role when the show was later adapted to film. His name graced the billboards of New York's Palace Theatre and resonated in the echoes of Broadway and radio waves. He became a beloved figure of entertainment during the heyday of the 1920s and 1930s. Known for his amiable and good-hearted comedic character, he often maintained a composed, impassive expression, serving as the serious counterpart in comedic pairings.

During his early career in vaudeville, Cook promoted his act as a fifteen-minute extravaganza showcasing a multitude of talents, including juggling, unicycling, performing magic tricks, hand balancing, playing ragtime piano and violin, dancing, rolling globes, wire walking, engaging in comedic dialogue and creating cartoons. He unquestionably delivered on his promises and enjoyed great acclaim during his prime, relying on his verbal performance, including the "Four Hawaiians" skit, to sustain his stardom.

Yet his star power did not eclipse his genuine love for Lake Hopatcong. After acquiring the Boulders cottage around 1924, Cook ingeniously transformed it into Sleepless Hollow, renowned for its ambiance and unique golf course. There, he built an extravagant estate. The estate had a playhouse, a mock railroad and an artificial lake, showcasing his love for whimsy and his roots in vaudeville. His fame never overshadowed his ties with the community; he integrated into local life, ensuring his children attended local schools and bestowing Christmas baskets on local children. Sleepless Hollow became a landmark of sorts in Lake Hopatcong and is a part of the history of the area.

Cook once confessed that despite his years on the lake, he had never successfully caught a fish. Nonetheless, he devised a clever strategy. Cook would affix a 150-pound plaster fish to his dock and raise it when

Joe Cook's piano at the Lake Hopatcong Historical Museum. *Author's collection.*

passenger boats passed by. This spectacle created the illusion that he was an accomplished angler. Additionally, Cook gained recognition for his renowned vessel, the *Bear Isle Bearcat,* which he christened *The Four Hawaiians* after his famous and uproarious stage performance. In this act, he would declare his intention to impersonate four Hawaiians but never actually fulfill that promise.

In a nutshell, Joe Cook was more than an exceptional entertainer; he was a beloved pillar of the Lake Hopatcong community, a figure whose enduring legacy continues to inspire. His affection for the lake permeated his performances, which served to cast Lake Hopatcong's luster far beyond its shores. As the Great Depression set in and vaudeville's popularity started to fade, Cook's career, like regions such as Lake Hopatcong, also began to suffer. His comedic style, which worked well on the stage, did not transition successfully into the age of sound films, often referred to as the "talkies." Consequently, his attempts to establish a film career were not successful. Cook lived at Lake Hopatcong until his death in 1959.

Another star at Lake Hopatcong was actress Lotta Crabtree. Born in New York City, Lotta's father, John, an Englishman, was swept away by the gold rush, and he left young Lotta and her mother, Mary Ann, behind. A year later, their transcontinental journey to reunite with him led them to a surprising revelation: John was nowhere to be found. In the heart of this unplanned turn of events, destiny wove an unusual path. The pair stumbled upon a jubilant troupe of entertainers, their lives awash with the vivid hues of performance art.

This fortuitous encounter proved to be a pivotal moment for Lotta. She was inspired to dance, sing and perform. Little did she know then that these skills she was mastering would soon catapult her to stardom, making her one of the most adored and highest-paid entertainers of her epoch before she found her home at Lake Hopatcong in 1886. She then indulged in the enchantment of Lake Hopatcong during countless summers that followed for a remarkable span of fifteen years. A true devotee of this picturesque retreat, she graced its shores as early as April, and her stay usually extended through the entire summer.

Eric Sloane, a notable painter and author, had a significant connection to Lake Hopatcong. In the expansive domain of Eric Sloane's illustrious career, he ventured into various seasonal occupations surrounding the lake, including his active involvement at local boatyards. In a remarkable display of generosity, Sloane graciously presented an intricately crafted and vivid depiction of the lake to Nick Steneck, the proprietor of Hockenjos Boat Company, in 1936. Over time, this masterpiece found its way into the possession of Nick Steneck Jr. and his wife, Peg. Recognizing the profound historical significance embedded within this artwork, the benevolent couple selflessly entrusted it to the Lake Hopatcong Historical Museum.

Eric Sloane's artistic prowess extended beyond his notable murals, which adorned both private residences and corporate establishments. A captivating photograph, captured between 1999 and 2000 by Wil Mauch, unveils a prominent Sloane mural situated within the headquarters of the now-defunct International Silver Company in Meriden, Connecticut. This arresting image takes its place within the comprehensive retrospective publication titled *Aware: A Retrospective of the Life and Work of Eric Sloane*, skillfully authored by Wil Mauch.

Enthusiasts of Eric Sloane's artistic pursuits may be astonished to discover his dedicated focus on ski scenes during the early 1950s. Eric's sister, Dorothy, recounted that his fascination with Switzerland's breathtaking landscapes was kindled by a voluminous "coffee table book" acquired in the late 1940s. This remarkable tome showcased vivid and saturated depictions of the Alps, captivating Eric's imagination. As a result,

Windlass Docks at Lake Hopatcong, by Sandi Astras. *Author's collection.*

Sangria at Lake Hopatcong, by Sandi Astras. *Author's collection.*

he embarked on a creative phase reminiscent of his renowned masterpiece titled *Sunlight and Shadow*. During this period, myriad paintings emerged, not only showcasing European settings but also encompassing the serene locales of New England and the rugged landscapes of the Rocky Mountain region in the Western United States.

In contemporary times, artists continue to draw inspiration from the scenic splendor of Lake Hopatcong. Some even set up their easels on the docks, captivated by the allure of the surroundings as they strive to immortalize the essence of this remarkable location through their art.

In the literary field, acclaimed author Rex Beach frequented the lake as a writing retreat. His prolific career includes over thirty novels, many of which were adapted into films. His prominence coincided with the zenith of the lake's popularity, particularly between 1911 and 1920. One of his notable works from this period, *The Iron Trail*, served as the basis for a 1921 silent film of the same name. The degree to which his lakeside residence influenced his novel remains speculative. However, the character Eliza Appleton from the novel comments, "I wanted the sensation. Writers have to live before they can write. I've worked the experience into my novel." Beach's fascination with exotic locales like Alaska and Panama contrasted starkly with his urban residence in New York City. Lake Hopatcong offered ample opportunities for adventure, aligning with his public persona.

The Iron Trail narrates the story of Murray O'Neil's endeavor to construct a railroad in Alaska to stimulate travel and commerce—an effect akin to the railroad's impact on Lake Hopatcong. Although this is speculative, Beach's plot insinuates a connection to Lake Hopatcong figures such as Hudson Maxim. The protagonist faces myriad challenges, including government interference, corrupt competition and challenging terrain. Beach consistently portrayed individuals conquering nature as heroes, providing a lens to explore his broader philosophies concerning the relationship between people and nature.

THE ARRIVAL OF HUDSON MAXIM AT LAKE HOPATCONG

While Maxim initially arrived at Lake Hopatcong to pursue his work on explosives, his involvement in the area expanded beyond his original intent. He stated, "It wasn't the beauty of the spot that attracted me. It was the powder works, and if they'd been in a desert, that wouldn't have deterred

Hudson Maxim's house and property on Lake Hopatcong. *Courtesy of the Hagley Museum and Library.*

me." However, Maxim's multifaceted interests led him to explore other pursuits at the lake. After residing in rented cottages for three years, he acquired land to establish his permanent residence. He built a laboratory, library and workshop, and he also developed his work residency.

As Maxim became more settled, expanding his laboratory and hiring assistants, he acquired more property and became "the owner of three-fourths of the land in the borough," according to his biography. To go along with his new home and property, Maxim built a stone boathouse and connected it to his house "by a high-arched causeway." He used the boathouse to store his boats and host parties, often inviting other famous inventors and engineers to join in the festivities.

Maxim's ascent from a proprietor of a motorboat, which was prone to frequent mechanical failures, as it often "stopped dead," to being the owner of the award-winning vessel named *Dreadnaught* marked a significant upgrade in his maritime pursuits. This progression was symbolic of his journey in the Lake Hopatcong community. From his humble beginnings in leased cottages, Maxim had effectively established himself, both materially and

This page, top: Hudson Maxim's Stone Boat House with Archway on the Left. *Courtesy of the Hagley Museum and Library.*

This page, bottom: A boat (possibly a Dreadnaught) heading toward Maxim's boathouse. *Courtesy of the Hagley Museum and Library.*

Opposite: Missiles in Maxim's home. *Courtesy of the Hagley Museum and Library.*

recreationally, within the community. His property holdings and nautical interests, reflecting his social and economic standing, demonstrated his successful integration into the lakefront society.

Possessing most of the land within the Hopatcong Borough, Maxim not only became the most consequential and committed resident, but he also assumed an active role in addressing the prominent lake issues of his era, primarily those affiliated with the Morris Canal. Clifton Johnson, Hudson

Maxim's biographer, depicted him as an "eccentric scientist," a figure of occasional ridicule, yet an inventive genius and profound philosopher of notable stature. One morning in 1923, Maxim shared his restless night's dream in which he was incessantly filling the Morris Canal with a truck, thus transforming MacCulloch's dream into his own nightmare. For Maxim, his engagement with the lake was not merely a pursuit of material gain; it was rooted in a deep-seated passion.

The Hudson Maxim Era

Perhaps the early twentieth century was, in fact, more representative of the influence of Hudson Maxim and, in his own way, his splendor. However, his main goal was to acquire land, a valued commodity, as explained in a letter at the Hagley Center in box 2, folder 15, accession 2147:

> *Land is the one common source from which all blessings flow. Everything, in the last analysis, harks back to the land. The whole wealth of the world is made in the manipulation of what is grubbed out of the earth.*

The farmer and the miner are the real lords and masters whom all must serve. All industrial arts and sciences busy themselves in the production of what the farmer wants, and constitute a body of consumers for the purchase of the output of the farms and mines.

It is undeniable that land held a pivotal role for Maxim. It served as the foundation for his wealth and resources that shape our world.

The local news was often punctuated with Maxim's activities and achievements as he grew his influence and land mass. For instance, a piece titled "An Interesting Trip," published in the *Breeze* on July 31, 1912, accentuates Maxim's societal standing, as he was a guest of "Mr. Fred E. Wadsworth, the multi-millionaire boat and engine manufacturer and his wife." The article found at the Hagley Center states:

The Ford Company intends bringing out a six-cylinder touring car for 1912. The car will have many new and unique features, and will be sold for the small sum of $900. Mr. Maxim bought the first one of these cars, Mr. Ingersoll the second, and Mr. Wadsworth, the third, the cars to be delivered early in 1914.

This innovative vehicle, the Ford Model T, was not just an upgrade for Maxim; owning a car would play a pivotal role in facilitating his access to Lake Hopatcong:

The Ford Model T was produced by Henry Ford's Ford Motor Company from October 1, 1908 to May 27, 1927. It is generally regarded as the first affordable automobile, the car that opened travel to the common middle-class American; some of this was because of Ford's efficient fabrication, including assembly line production instead of individual hand crafting.

However, the Ford Model T, when purchased by eminent figures such as Maxim, became more than a mere means of transportation; it evolved into a symbol, vouching for the booming automobile-centric ethos of the era. For Hudson Maxim specifically, the acquisition of this innovative automobile was not merely a practical transaction, but a strategic move that allowed him to craft an intricate web of impact and possibility throughout Lake Hopatcong. This wasn't solely about the utility of a personal vehicle but was intertwined with a larger narrative of a booming time. The automobile, especially one as emblematic as the Model T, served to bridge disparate

Hudson Maxim's car. *Courtesy of the Hagley Museum and Library.*

geographical landscapes. This enabled a smoother, more integrated blend of interactions and exchanges. The Model T was an invention that shaped many regions, including Lake Hopatcong.

Beyond the one example above, Maxim acted and innovated in ways that truly represented the era, establishing his reputation at Lake Hopatcong during a time of significant change and development. Essentially, Hudson Maxim's diverse and notable efforts in various fields established him as a highly influential person at Lake Hopatcong in the early twentieth century. His impact was both wide-ranging and profound, shaping various aspects of society from technology and warfare to literature and transportation. Thus, his pervasive influence, combined with his active involvement and transformative work in Lake Hopatcong, strongly supports the assertion that this period was indeed mostly the "Hudson Maxim era." This term aptly encapsulates his influence, underscoring his instrumental role in shaping the social and cultural landscape of the time.

As Maxim's prolific presence grew, Lake Hopatcong not only flourished as a resort destination, but it also established itself as a stable community replete with resources. A significant event transpired in August 1911, as reported by the *Lake Hopatcong Breeze*: a free public library opened in the

Hudson Maxim enjoying recreational time outside of the laboratory. *Courtesy of the Hagley Museum and Library.*

neighboring Mount Arlington. Community members eagerly converged to acquire their books, with over five hundred displayed in meticulously crafted bookcases that greeted visitors upon entry. For the local population, the library's inauguration signified the commencement of a new epoch in literacy and education, transforming the region beyond merely a vacation spot. The library embodied the community's progressive spirit by serving as a beacon of development and growth, offering convenient access to an extensive variety of books and facilitating knowledge expansion.

In 1911, the Hopatcong Rifle Clubs sought membership in the NRA, the Boy Scouts considered organizing a branch and even a casino proposal emerged. The lake experience thrived, boasting unique features, such as a store with a canal entrance, allowing customers to make purchases without disembarking their boats—the only of its kind. This period witnessed the evolution of a robust community with exciting and innovative establishments arising.

"At the Lake Is Just About Ideal"

Maxim, captivated with the lake environment, frequently indulged in creative liberties when portraying climatic conditions. At times diverging from science and empirical evidence, Maxim's depictions of weather and the environment were rooted in hyperbole. This tendency is evident in his speech delivered before the canal commission on November 16, 1912, in which Maxim redirected attention from business matters to emphasize the alluring climate. During this discourse, he lauded the alleged therapeutic attributes of Lake Hopatcong. Found in box 1, folder 18, accession 2147, he wrote:

> *At this moment, every resident of Lake Hopatcong is recollecting many such cases, and there are individuals present here tonight who owe their continued existence to the life-saving grace of Lake Hopatcong.*

He went on to say:

> *Three years ago, a certain literary gentleman, who is now our acting Borough Clerk, and one of us this evening, was a pale, emaciated consumptive, in the pent airs and disease-laden whirl of New York City, and one day he fell in the street, stricken with that premonitor of death, the pulmonary hemorrhage.*

Of this "literary gentleman," Maxim claimed that "in less than six months, he was well again, and fat and sound, and not a germ of tuberculosis to be found by the most searching diagnosis." Such rhetoric was characteristic of Maxim's prose when discussing the qualities of the lake.

This embellished language was not just a prominent characteristic of Maxim's when extolling the attributes of the lake. The climatic conditions of Lake Hopatcong were frequently highlighted by other authors. People endeavored to portray the natural locale as a sanctuary of temperate weather, often prioritizing promotional intent over factual accuracy. The *Illustrated Guide to Lake Hopatcong for the Season of 1898* went so far as to extol the virtues of the lake during winter, traditionally a period of sluggish or absent tourism. The *Illustrated Guide* stated:

> *The summer visitors of the lake, no doubt in the winter as they gather around their cosy* [sic] *fire-places in New York and other cities, occasionally*

> *think of Lake Hopatcong, and probably some of them look upon it about as they do Kamskatka; as a cold and frozen region of snow and ice, without inhabitants, and think some venturesome person like Lieut. Peary ought to be commissioned to come up here and view the arctic wonders of the place, and perhaps dig out any inhabitants who had ventured to stay here too late.*

Contrarily, the guide reassured readers that Lake Hopatcong during winter was "quite animated" and the weather "at the lake is just about ideal." It even made a spurious assertion that residents of Lake Hopatcong were unaware of cold fronts until they received news of them from New York.

With Lake Hopatcong at an elevation of 950 feet above sea level, compared to New York City's 33 feet, the prospect of milder winters at Lake Hopatcong is an unrealistic portrayal. This discrepancy, despite historically colder winters at Lake Hopatcong, was supposedly due to the air being "invigorating, but not harsh or piercing" according to the *Illustrated Guide*. The region has experienced warming trends throughout the twentieth century and into the twenty-first century, yet even today, it endures colder winters than New York City, accompanied by more snowfall. From personal experience, this author can affirm that even with contemporary snow shovels and pet-friendly ice melts, he finds himself snowbound more frequently than his neighbors nearer to New York City.

Despite the lake's lack of appeal for recreational visitors during winter, it fostered a thriving industry due to the very frigid climate that the guide romanticized. Half a decade following the initiation of the railroad, the Brady siblings set foot at the lakeside and inaugurated an ice venture. Throughout the winter season, they would extract large frozen portions from the lake, leading Lake Hopatcong to become a renowned hub of the organic ice industry in the eastern part of the country for an extended period, concluding around 1930. The chilly months saw over one thousand men actively employed, and the ice-loaded cargo trains surpassed the summer passenger trains in length.

Mountain Ice Co., the most substantial ice company in the region, concluded its lake ice collection in 1934. In the subsequent five years, one of the nation's largest ice storage facilities, standing fifty-six feet tall was torn down, symbolizing the cessation of extensive ice gathering operations in New Jersey. The entire procedure of ice collection and its storage was both thrilling and chilling. At first, the ice businesses recruited workers from New York, eventually shifting to local labor and people brought in from Maine. The industry thrived because in the early twentieth century, the lake would

Cracked winter sand at Hopatcong State Park. *Author's collection.*

frequently freeze during the winter, providing a substantial resource for ice. During the winter, ice laborers would meticulously chisel out blocks up to two feet in thickness. They would then float these sizeable ice chunks toward massive ice storage buildings, cover them with sawdust and then load them onto railway carriages destined for the iceboxes of New York City and farther

Shadows on Lake Hopatcong ice. *Author's collection.*

places. The 1898 *Guide* proudly described how "the general exuberance and bustle create a lively scene." To put it in perspective, it explains:

> *If the cakes of ice as they cut them, about two feet square, were placed in a single row, stretching toward the east, they would cross the Ocean, go through Europe, Asia, back over the Pacific to San Francisco, and then three thousand miles across to Lake Hopatcong again, about sixteen or eighteen thousand miles at this latitude, and there would be ice to spare at that.*

This was prior to a century of rising temperatures. The conditions surrounding the lake differed significantly back in the twentieth century.

The lake now freezes less often, but there are still myriad activities attracting people who come together to partake in the enchanting winter wonderland. While the lake is not useful for ice commerce, it has evolved into a gathering place for families, providing an opportunity for activities like ice fishing, walking and skating. As the ice thickness increases, individuals cautiously venture onto the frozen surface, while bundled-up children joyfully slide around, supervised by their attentive parents.

The lake is still a sanctuary where individuals can momentarily escape the burdens of their everyday lives and simply revel in the brilliance of the moment. The emergence of drone technology has gained popularity, finding a fitting setting for operation on the lake. The resounding hum of the propellers and the buzzing of drones fill the air as enthusiasts seize the opportunity to capture aerial footage of the icy landscape. Some even bring picnics, setting up small tables and chairs on the ice and braving the cold to relish the refreshing winter air. It is a changed environment but one worth preserving and savoring.

GUARDIAN OF THE LAKE

Maxim highlights the unique characteristics of Lake Hopatcong, emphasizing that due to its irregular shape, the lake can support a shore population equivalent to that of six round lakes of the same area or even a single round lake that is thirty times larger. However, he also raises concerns about the lake's vulnerability to pollution due to the relatively small size of its watershed, explaining that the water of Lake Hopatcong becomes twelve times more polluted when subjected to an equal amount of shore population. Maxim's insightful analysis and warning about the potential consequences of overdevelopment underscore his proactive approach in recognizing and addressing future challenges faced by Lake Hopatcong. This foresight regarding the adverse impact of overdevelopment on the lake's ecosystem remains pertinent to the present day.

In addition to recognizing the significance of the lake community in maintaining the health of Lake Hopatcong, Maxim actively participated in the endeavor to shape the updated infrastructure. The *Breeze*, on June 26, 1912, attests to his involvement, stating:

> *Maxim Drive is now nearly completed for about a mile of its length. This fine roadway will extend from the present main road a short way above the River Styx Bridge through to the extreme end of Maxim Park, where it will connect with a road to be built by the Hopatcong Park and Bear Pond Land Company and also the road that is now in course of construction by Byram*

The Lake Hopatcong Watershed. *Author's collection.*

> *Township. Maxim Drive will be nearly two miles long, and will follow the Lake a few hundred feet back from the shore. It will be a beautiful Lake view drive. Mr. Maxim is having all of the undergrowth and useless scrub growth cleared by a scientific forester from all the shore-front lots between Maxim Drive and the Lake.... When this is done, there will be a road clear around the Lake, and it will be one of the most beautiful automobile drives to be found anywhere in the eastern states.*

This notable contribution played a pivotal role in the transformative process of the still-developing region. Maxim Drive borders where the chestnut trees would have blocked the view of the lake. While the article does not explicitly state this, it does show how the trees were perhaps a barrier for the landscape that Maxim envisioned, since the trees would have cut off the view of the lake from Maxim Drive.

Maxim facilitated the cutting down of some lakefront chestnut trees, believing they were unsalvageable due to insect infestations. This action elicited at least one complaint. On July 27, 1912, the *Lake Hopatcong Breeze* published an article titled "Mosquitos and Chestnut Trees" (box 1 folder 18 accession 2147), in which F.G. Himpler, the president of the Mount Arlington Protective and Improvement Association, criticized the tree-cutting:

> *The chestnut trees are very valuable trees for our summer resort. I do not refer to the fruit they produce, but to the protection they give us against mosquitos. I observed years since and especially again this year, that if we had mosquitos in the spring, they disappeared as soon as the chestnut trees began to bloom the beginning of July. It seems the odor of the blooming trees is unbearable to the mosquitos (the same, as for instance the odor of citronella) and drives them to other localities.*

Himpler endeavored to preserve the lake's character, and as the head of the protective agency, he possessed a professional perspective centered on utilizing the chestnut trees to repel mosquitoes.

Maxim's involvement in the tree-cutting remains unclear, but he did not defer to Himpler's expertise. On August 5, Maxim countered Himpler's argument:

> *Now in the first place, it is not an insect at all, that is attacking the chestnut trees, but it is a far more deadly thing. The chestnut blight is due to a fungus growth which attacks the trees between the bark and the sap. Hence it is called the chestnut disease; sometimes chestnut canker; sometimes, chestnut cancer. Its scientific name is* Castanea dentata *(Borkh) or* Diaporthe parasitica *(Murrill). The disease rapidly spreads under the bark, soon girdling a branch or the body of the tree itself, causing the immediate death of the branch or tree above the infection.*

Maxim's understanding of the situation appeared more comprehensive than Himpler's. Maxim was not just a resident; he was a discerning observer who demonstrated a deep interest and knowledge of the area and often utilized his scientific expertise. The story unfolded in a dichotomous manner, with Himpler perceiving Maxim as a destructive force, whereas Maxim considered himself a custodian of the regional ecosystem's holistic wellbeing.

Aligned with his commitment to champion various causes aimed at regional development, Maxim actively engaged in a battle with the post office to enforce efficiency, viewing it as a crucial component within his overall vision for the area. Employing a favored tactic reminiscent of his equivalent to contemporary social media platforms, Maxim composed a letter titled "Confusion of Post-Offices on Lake Hopatcong" (box 1, folder 18, accession 2147), presumably intended for publication in a local Lake Hopatcong newspaper. This letter delved into the intricate complexities surrounding the local post offices and highlighted their inefficiencies within the Lake Hopatcong region.

Presently, this issue remains relevant, as Lake Hopatcong spans across both Sussex and Morris Counties, encompassing four distinct boroughs: Hopatcong (Maxim's place of residence), Jefferson Township (which includes the town of Lake Hopatcong), Mount Arlington and Roxbury Township. It is worth noting that the designation "Lake Hopatcong" pertains not only to the body of water itself but also to the aforementioned town situated within the borough of Jefferson Township.

During Maxim's era and continuing into the present day, the issue of postal confusion persisted. Maxim's unwavering dedication extended to even the minutest aspects of town affairs, leading him to undertake efforts to streamline the postal system surrounding Lake Hopatcong. In an assessment that may be perceived as biased due to the writer's Hopatcong affiliation, Maxim (box 1, folder 18, accession 2147) accurately summarized, "It is doubtful if there is any other place in America where so much difficulty is experienced in getting one's mail and express directed to its proper office." He voiced his dissatisfaction, expressing:

> *The fact that he is known to live on Lake Hopatcong results in a large percentage of his mail being addressed Lake Hopatcong, and it has to be forward to him to his proper post-office. If one lives in the Borough of Hopatcong, his business associates, friends and acquaintances, when writing him in order to be sure that he gets his mail promptly, and knowing that he lives on Lake Hopatcong, prefix the word Hopatcong with Lake, for safety, and his mail therefore does not come promptly.*

Ironically, when Maxim died, there was confusion in accurately identifying the specific municipalities involved. *Time* magazine erroneously reported that his death occurred "at his home at Maxim Park, Lake Hopatcong, N.J.," when in reality, it took place in Hopatcong.

As part of his lake pride, Maxim published a poem authored by his house guest Paul West in the *Breeze* newspaper in 1912. This poem (box 1, folder 18, accession 2147) aimed to present the lake through the perspective of an outsider, providing a unique glimpse into its splendor:

> *Oh, the Lake was like a silver streak that gleamed along the side!*
> *The sky was one long blur of blue, with azure glorified.*
> *And the scenery! Oh, the scenery—*

A successive gob of greenery!
And my evil past sped by me
With its terrors to defy me!
And it took two men to watch us speeding faster than the crows—
One to say, "Hi! Boys, she's coming!" and the other, "There she goes":
Paul Revere was never in it,
Not with us, one single minute,
On that speeding, rushing, roaring, ripping, whizzing, whirring day,
When we motored out with Maxim on the swiftest fifth of May!

Significantly, Maxim assumes the role of a captain within the poem, aligning with his self-perception as a key figure, akin to a discoverer in terms of his envisioned legacy. This portrayal not only validates Maxim but also serves to elevate the extraordinary qualities of the lake as depicted in the poem.

The Struggle for Lake Hopatcong's Future

Maxim's notable and influential contribution lies in his endeavor to liberate the lake from the clutches of the government. The canal was facing the end of its run, and the State of New Jersey proposed turning the lake into a reservoir to provide water for the state. By 1902, the canal's shipment of coal had plummeted to a mere 20,411 tons, a significant decline from its zenith of 459,175 tons in 1860. Although the canal was not officially abandoned until 1924, plans for its abandonment were already underway, and New Jersey was initiating preparations to transform the lake into a reservoir.

The reservoir project held significant importance for New Jersey, as it aimed to provide water to major cities such as Jersey City and Newark. However, designating the lake as a reservoir came with the consequence of prohibiting swimming, fishing, boating and other water-based activities. This would cause property values around the lake to plummet and would diminish tourist interest in the area. Maxim contended that the rights to the lake should revert to the landowners around the lake. As tensions escalated over the reservoir proposal, Maxim found himself faced with the dissolution of his aspirations to shape the lake according to his own vision. This led him to embark on a literary battle.

To defend it as a recreational lake, Maxim composed a text in 1913 titled *Lake Hopatcong the Beautiful: A Plea for Its Dedication as a Public Park and for Its Preservation as a Pleasure and Health Resort for the Benefit of All the People.*

Throughout his book, Maxim employed a series of rhetorical strategies to accentuate the significance of the lake. To foster a sense of pride in the lake, he drew parallels between Lake Hopatcong and Niagara Falls, arguing that "every citizen in the State of New Jersey is proud of Hopatcong." He extolled the lake as a place of delight and renewal, stating, "Everyone who has ever visited it is enthusiastic about Hopatcong; and thousands of New Jersey citizens enjoy its beauties and are renewed in health there every year." With a touch of dramatic flair, he presented the canal as a danger to the "queen of all the lakes of New Jersey," who was "on trial for her life."

Recognizing that those residing farther from the lake might be less invested in the cause, Maxim provided a compelling analogy, suggesting, "[T]he inhabitants of Mars, being farther away, do not get so much of the sun's warmth as we do, still their interest in the sun is as actual as ours." This statement underscored the universal relevance of a clean lake, irrespective of one's proximity to it. As a shrewd businessman and a poet, Maxim also expounded on the economic ramifications of the lake: "Each automobile party spends, on the average, more than fifteen dollars a day." Despite Lake Hopatcong never attaining the touristic prominence of Lake George or Niagara Falls, it remained Maxim's cherished home and a site of immense personal significance.

While Maxim confessed his vested interest in the lake, owing to the six hundred acres he owned, he ardently proposed that the issue transcended personal interests and was, in fact, a collective responsibility. Maxim shared his preservation agenda, which centered on his ambition to preserve the lake for recreational activities, such as boating, swimming and residential pursuits. Urging the populace to rebel against any governmental plans, Maxim appealed, "Now, it is your duty to yourself and to your family, and to every friend and neighbor of yours, that you should lend a helping voice to prevent you and them from being robbed." Furthermore, he contended that a reservoir will result in the destruction of "the most beautiful lake and mountain resort of New Jersey."

THE LAKE HOPATCONG CORPORATION

In his endeavor to advance local objectives, Maxim assumed a central and influential role in the establishment of the Lake Hopatcong Corporation circa 1920 that was created to preserve the lake. Despite occupying the honorary position of vice-president within the corporation, he assumed a leadership role in driving the organization's initiatives to protect the lake from the proposed reservoir project. This is substantiated by the exchange of correspondence between Maxim and Louis Schwab, the president of the Lake Hopatcong Corporation. Notably, on July 22, 1921, Maxim expressed his grievances to Schwab (box 1, folder 19, accession 2147), articulating:

> *I looked for you all day yesterday to call upon me, but you failed to put in any appearance. I called to you from my boat some time ago while passing your dock and asked you if you could not come round and see me, to talk over Corporation matters.*

Maxim conveyed to Schwab that although he was occupied with various commitments, he expressed a desire to meet with him prior to the upcoming annual meeting of the Lake Hopatcong Corporation. In light of the substantial workload apparent from his archives, Maxim aptly recognized his own contributions, stating, "I have done all that one man could do in the interest of the Corporation since the last meeting and I am still doing all that can be done." This acknowledgment underscores the extent of Maxim's dedication and the breadth of his endeavors on behalf of the corporation.

Despite being the unequivocal leader, the reasons behind Maxim's decision not to assume the position of president within the Lake Hopatcong Corporation remain unclear. He contributed significant time and energy and owned eleven times more stock than any other member of the corporation. One possible explanation could be the scrutiny he faced during past canal abandonment meetings in Trenton, New Jersey, where individuals questioned his motivations as the largest landowner around the lake. Nevertheless, Schwab, acknowledging Maxim's influential role, frequently deferred to Maxim's unofficial authority. In a letter dated July 26, 1921, Schwab provided an elaborate explanation for his inability to meet with Maxim, citing his responsibilities in relieving his wife by attending to the children's needs throughout the day and assuming the duty of bidding them goodnight before retiring to bed himself. This pattern of excuses persisted, as Schwab openly confessed:

> *Last Thursday, however, I did stay up and so had planned to call to see you, but my time was so taken up with necessary duties and I will confess a stealing away for an hour and a half for golf, so that recollection of my promise to call failed to connect with a possible time of doing so.*

Maxim was chasing the president, who appeared to be shirking his responsibilities.

Schwab was not the sole individual who regarded Maxim as a leader within the Lake Hopatcong Corporation. Elmer King, who later assumed the presidency of the corporation in 1922, also seemed to yield to Maxim's authority. In a letter dated December 14, 1920, King expressed his apologies for missing Maxim's meeting, citing his engagement with a distressed individual facing property foreclosure who sought his assistance. King provided a detailed explanation and pledged to attend the subsequent meeting, exhibiting a demeanor akin to that of an employee addressing a superior. Numerous letters of a similar nature were directed toward Maxim, further exemplifying his prominent role within the corporation.

Despite the absence of a clearly defined power structure, the Lake Hopatcong Corporation demonstrated remarkable resourcefulness in devising a strategic plan against the reservoir idea. It primarily revolved around some so-called Niles Rights. These rights, with origins dating as far back as 1715, granted ownership of the land beneath the lake as it existed during that colonial era. They had been acquired by various private individuals. Historical records show that these lands were initially obtained directly from King George I. His early interest in the region surrounding the lake was primarily focused on its timber resources rather than the lake itself. Over time, the rights to the land beneath Lake Hopatcong eventually passed into the possession of Lord John Berkeley and Sir George Carteret. They were prominent figures in the colonial history of New Jersey. They were among the eight Lords Proprietors who were granted ownership of the Province of New Jersey by King Charles II of England in 1664.

In 1882, there was a significant turning point in this narrative. During that period, the current stakeholders of the Niles Rights offered any unclaimed uplands within 300 feet of the high-water mark at a public auction. The highest bidder for this valuable property was Nathaniel Niles, a resident of Newark and Madison. He was also a lawyer and an assemblyman in New Jersey from 1871 to 1872 who held influence and financial means. With the acquisition of these substantial rights, Niles subsequently made a series of transactions. He sold sixteen-fiftieths of his interest to two different parties, and later, he parted with the remaining eighteen-fiftieths to a third party. This intricate web of ownership signaled a fragmented landscape of ownership. The corporation consolidated and acquired the entirety of these rights, thereby gaining full control over the destiny of the lake. This tactical move granted them the authority to influence the development and

Nighttime at Lake Hopatcong. *Author's collection.*

management of the lake area in a comprehensive manner, shaping its future as a vital resource and recreational destination.

Maxim diligently cut out and preserved a newspaper clipping that effectively outlined the strategic approach. The newspaper clipping, dated August 18 (box 1, folder 18, accession 2147), succinctly encapsulates the historical context of these rights, providing a comprehensive summary:

> *To ensure a solid front of property owners holding land here affected by the Niles rights against any attempt to obtain the waters of Lake Hopatcong for a potable supply, the Lake Hopatcong Corporation, consisting of a score of residents, has obtained confirmatory rights at a cost of about $30,000.*

> *The rights arise from deeds acquired from Nathaniel Niles in 1882 for all allotted land under water and 300 feet above high water mark at Hopatcong and Culver's and Swartswood lakes....Mr. Maxim and others associated with him then purchased the Kinney and Ward rights for $10,000 each. He had previously given $10,000 for his original share. The original deed to Niles was made by Charles E. Noble of Morristown, as president of the board of East Jersey Proprietors, who acquired their rights from the King of England by royal grant.*

The blueprint of the corporation revolved around getting these deeds.

To make their acquired limnological property legal, the corporation had to find a precedent. They cited *Cobb v. Davenport, 32 N.J. J.L. 369* (a New Jersey case from 1867). The case was kept in Maxim's papers at the Hagley Center, and it argued:

> *The policy of the common law is to assign to everything capable of ownership a certain and determinate owner, and for the preservation of peace, and the security of society, to mark, by certain indicia, not only the boundaries of such separate ownership, but the line of demarcation between rights with are held by the public in common, and private right.*

It established that anything could be owned and set criteria through:

> [A] *test by which to determine whether waters are public or private is the ebb and flow of the tide. Waters in which the tide ebbs and flows, so far only as the sea flows and reflows, are public waters; and those in which there is no ebb and flow of the tide are public waters.*

To establish the scope of ownership for this legal test, the corporation conducted a survey "showing a return of 1,231 acres affected by the Niles rights."

Interestingly, Maxim regarded the Niles Rights as more of a burden than a privilege. He was upset about a tax bill that amounted to $9.17 owed to the Borough of Hopatcong and an additional $46.80 owed to Byron Township. While not an excessive sum, it symbolized an extra responsibility that Maxim was reluctant to accept. Never one to remain silent, he expressed his dissatisfaction in a letter addressed to the editor of the *Sussex Register*, making his grievances public:

> *All of the open lake included within the Niles Rights and now owned by the Lake Hopatcong Corporation is being freely used by the public without any restrictions or charge for its use being imposed by the Corporation, and I understand it is not the intention of the Corporation ever to charge the public anything for the use of the open lake.*

Maxim felt that since the lake was open to the public, the land covered by the Niles Rights lines should be exempt from taxes (box 1, folder 20, accession 2147). However, in the end, Maxim still had to pay his bill. This was because, despite being accessible to the public, the land beneath the lake continued to be privately owned, which was the very purpose of having the rights. Consequently, having taxable property raised questions about the future accessibility of the lake for the general public.

On August 25, 1921, Maxim wrote to Elmer King, addressing the significance of the Niles Rights and expressing reservations regarding these holdings. He recounted how he acquired eighteen-fiftieths of the Niles Rights for a sum of $20,000 in cash, but their value was already appreciating. Subsequently, as the financial holdings continued to grow, other shareholders within the corporation began expressing an interest in selling off their portions. Despite Maxim's initial reluctance, an agreement was reached that selling limited sections was a fair way for members to recover some of their costs.

While Maxim endorsed this decision, he adamantly opposed any profit-seeking motives among the corporation's members. He reminded the members of their initial obligation to utilize the Niles Rights to preserve Lake Hopatcong from a potential reservoir. Despite the fact that he estimated that the value of his shares had doubled, Maxim led by example and made it unequivocal that he had no intention of profiting from them. Maxim extended a generous offer to fully reimburse any member of the corporation who harbored fears of financial loss, providing them with both their principal investment and additional interest. Despite the consensus that was reached, he continued to correspond with King, expressing his disappointment with the motives of the other members.

In theory, the very existence of the rights posed a risk. Even if Maxim and the members of the Lake Hopatcong Corporation had the intention to open the lake for public use, there was no guarantee that any future rights holders would not impose restrictions to further their own interests, contrary to Maxim's wishes. The Alamac Hotel shared similar concerns, prompting its owner, Harry Latz, to take action. Latz dispatched a representative named Carlton Godfrey to investigate the matter, with Godfrey attending a meeting at Maxim's residence on September 10, 1921. Following the meeting, Godfrey prepared a report for Latz, outlining the outcomes of their discussions:

> *The writer does not question the good faith of any one of the directors or stockholders of the Lake Hopatcong Corporation, but the danger of a change in the board is ever present, and the possibility of making a profit upon such a small amount of capital would be tempting to some speculator who might acquire knowledge of the situation.*

According to Godfrey, Maxim himself proposed

> *a resolution embodying a declaration that "all of the lands under the waters of Lake Hopatcong acquired from Theodore A. Gessler excepting a strip of land extending from high water mark into the Lake fifty feet and excepting such conveyances as have been made therefrom by this corporation, be declared to be open to the free use of the public for boating, bathing, fishing and winter sports." This declaration was not adopted by the meeting for the reason that the officers and many of those present insisted that the effect of the adoption of such a declaration would be to reduce the value of the lands under water known as the Niles rights in the event of a condemnation of the Lake by the State of New Jersey or by authority of Acts of the Legislature.*

The proposal was rejected. Despite not having a formal proposal in place, Godfrey received assurances from the members that the lake would remain accessible to the public. Nevertheless, he expressed his perplexity, questioning why the Lake Hopatcong Corporation did not adopt the resolution proposed by Maxim so that there would be no doubt. In his report to Latz, Godfrey concluded that it was inconceivable for individuals to invest significant sums in properties around the lake without ensuring their right to freely enjoy the lake for recreational purposes. This highlighted tensions between the members of the Lake Hopatcong Corporation and business owners who relied on the lake's waters for commercial gain.

The corporation's decision to vote against the proposal might not have stemmed from selfish motives. Instead, it was rooted in a misalignment with the corporation's mission. They argued that by preserving the existing rights, valued at around $3 million, without imposing any restrictions, they could prevent the State of New Jersey from acquiring the lands through any form of eminent domain. The substantial value of these holdings protected against such an acquisition.

Nonetheless, Maxim remained unwavering in his dedication to his plan. It wasn't a superficial attempt to please Godfrey; it emerged from his genuine belief that this was the right course of action. This stance led to a division between him and the dissenting members of the group. Expressing his dismay, Maxim penned a letter to T. Elliot Tolson, the commodore of the corporation on December 19, 1921, in which lamented:

> *It would have been far better surely for those in control of the Corporation to have devoted their time and energies to make sales last summer and fall, rather than to working against me both inside and outside of the Corporation. Something worthwhile might have been done before the people left the Lake for the winter, but now there is little opportunity to make sales.*

Maxim's once formidable influence had significantly waned. This diminished role is evident through his diminishing capacity to impose his wishes on the corporation. He now found himself hoping that certain members would choose to sell and depart rather than obstruct his path.

During the following year, Maxim's influence was further weakened by a significant shift within the board. The board was reconstituted in an August 8, 1922 meeting. William E. King became the new president, while several members of the board, including Tolson, who had grown to be a trusted confidant of Maxim, left. At this same meeting, property owners expressed their strong interest in obtaining the Niles rights to their affected lands, an idea that was supported by Maxim. Already feeling neglected, Maxim now faced the challenge of engaging new board members and a new public opinion. The confluence of these events altered the political landscape around the lake.

Feeling unheard, landowners took decisive action against the Lake Hopatcong Corporation's land claims. An article published in the *Lake Hopatcong Breeze* on October 28, 1922, reported on a collective lawsuit aimed at regaining ownership of the lands encompassed by the Niles Rights. The residents were perplexed by the validity of the maps used by the Lake Hopatcong Corporation to assert its rights over the Niles Rights territory. The *Lake Hopatcong Breeze* conveyed these concerns, emphasizing that, "From inquiries made, however, the *Breeze* has reason to believe that in many instances the surveys as depicted on the map are incorrect and that the Lake Hopatcong Corporation will have great difficulty in proving its claim to ownership." This public expression of doubt not only questioned the accuracy of the corporation's maps but also cast doubt on its established legal standing, which had previously been upheld in the case of *Cobb v. Davenport*.

With the removal of the reservoir threat, along with growing legal challenges and negative public sentiment, the corporation decided to surrender the Niles Rights. The historical records around the financial particulars of these transactions are ambiguous. This action guaranteed the lake's preservation as a public waterway, accessible to everyone, free from any state or private interests. As a result, having accomplished its goals, the Lake Hopatcong Corporation ceased to have a meaningful purpose and was subsequently dissolved.

Maxim's Legacy

Despite the successful conclusion of the battle over the lake, Maxim continued to advocate for the lake until his passing in 1927. His legacy is multifaceted, encompassing both commendable aspects and instances of stubbornness and defensiveness. Maxim drew parallels to the sensibilities of battle in his writings, illustrating his complex perspective on various issues. This is exemplified by his mention of a soldier pardoned by Abraham Lincoln. In Maxim's own words (box 2, folder 12, accession 2147):

> *Sometimes emotions of surprise, or even those of love, will force the duty of the heart till thought and purpose fade and fail in the blood-burdened brain, and infant weakness takes the helm of reason; or fear may paralyze the heart and rob the brain, and judgement, fainting, fall.*

While Maxim had his shortcomings, such as his stubbornness, they were born from his passion for the lake.

After the culmination of his battles, Maxim directed his attention toward other endeavors that stirred his passions. Preserved within his archives is an article from an unidentified Newark newspaper that was found in his papers. It is dated August 18, 1922, and it outlines another ambitious undertaking. Contradicting his earlier concerns about overpopulation, Maxim expressed his desire to transform the surrounding area of the lake into a city. The headline of the article aptly captured the essence of Maxim's legacy: "Hudson Maxim, Who Put 'Hop' in Hopatcong, Wants Another North Jersey City." Acknowledging Maxim's deep connection to Lake Hopatcong, the paper affirmed that "Hudson Maxim is as much a part of Lake Hopatcong as Nolan's Point or the Alamac."

Furthermore, the article accurately portrayed Maxim as an unwavering champion for causes he embraced, asserting that "when [Maxim] takes up

A tribute to Hudson Maxim at Maxim Glenn Park. *Author's collection.*

the cudgels for a cause, that cause will find no more ardent champion." While the article is reminiscent of an obituary, albeit for his influence, its writer likened the significance of Maxim's work to the groundbreaking discovery of atoms in 1889, suggesting that Maxim challenged and disproved of the ideas put forth by other scientists. However, the writer dismissed Maxim's idea of a new city as a wasteful expenditure. Over time, Maxim's power—or, at the very least, his position—gradually waned, and he was no longer at the zenith of his influence.

The motivations behind Maxim's exceptionally active role in Lake Hopatcong, compared to other notable residents in the area, remain somewhat elusive. Among his collection of papers at the Hagley Center is a manuscript bearing the annotation of 1907 (box 2, folder 30, accession 21). Titled "Old Nance of the Underworld," the piece was written by an author identified as Miscell Ibidanon. It is worth noting that Ibidanon appears to be a pseudonym, as no concrete evidence of an author by that name from the corresponding period could be found. The given first name seems to be an abbreviation of "miscellaneous," while the surname is a combination of the Latin term *ibidem* (meaning "in the same place") and the abbreviation "anon" (indicating anonymity).

The manuscript recounts the experiences of an elderly woman who arrives in a New England town, featuring a scene in which a baby becomes trapped in a house fire. In this passage, Ibidanon writes, "Old Nance, turning upon the mob, cried, 'Are not some of you men going to save that baby? Are you going to let that child roast up there? Cowards!' She turned and rushed into the hall. No-one tried to stay her." Despite her desperate pleas for assistance, the bystanders not only fail to rescue the baby but actively impede her rescue efforts. This passage conveys a poignant message that can provide insight into the author's mindset. It is conceivable that Maxim saw himself as a custodian of the lake, driven by a sense of duty due to the lack of engagement from others in championing the cause.

The narrative further unravels, offering a glimpse into the father's profound sense of powerlessness as he desperately seeks assistance from others to rescue his own child. A significant passage reads:

> *John Calvin Small stood wringing his hands, alternately praying and crying that he would give ten dollars reward,—fifty dollars—a hundred dollars,—to anyone who would save his child; while Mrs. Small, in hysterics, was held by some kindly persons, to prevent her rushing in and losing her own life in the endeavor to rescue her baby.*

The infant symbolizes the prospective trajectory of this family, while the father's mere lamentations and appeals for assistance expose his own passive stance.

Only Old Nance emerges as the rescuer, while the rest of the onlookers remain idle. The parallels between the literary work and reality cease at this point—with the exception of Old Nance receiving a ten-dollar reward from a neighbor, which Calvin, under the pretense of expressing concern for her injuries, pilfers from her possession. In this metaphorical portrayal, Ibidanon highlights the prevalent reluctance of individuals to fight for their own future, underscoring that even a solitary individual who undertakes this responsibility can effect meaningful change, despite facing criticism and eventual exploitation. Maxim perhaps sought to communicate his personal sentiments regarding the efforts he invested in preserving the future of Lake Hopatcong.

Perhaps the most enlightening way to truly grasp the essence of Hudson Maxim's mentality is to delve into this intriguing anecdote:

> *When Young Maxim was examined by the Committee to test his qualifications for the position of teacher he was asked no questions about his educational abilities, but was questioned regarding his ability to whip the bully, who had thrown out two or three previous school masters. Maxim said he was willing to undertake the job and asked the examiner to feel of* [sic] *his muscles. He was accepted immediately.*

This intriguing piece of information stems from an eleven-page manuscript that was unearthed among Maxim's documents housed at the Hagley Center. The document is inscribed with the date May 15, 1916. Though the authorship and exact timeline of the document remain obscure, a handwritten inscription along the top the first page catches the eye: "Given to Henry Irving Dodge." This note suggests Maxim had considered collaborating with Dodge, a celebrated American author of his time, before ultimately joining forces with Clifton Johnson. Interestingly, Maxim later recounts the episode of the bully from a personal viewpoint in his book with Johnson, *Reminiscences and Comments*.

Whether he was showing his muscles or using his pen, Maxim was rarely humble. On June 28, 1919, Maxim commemorated his twentieth anniversary at Lake Hopatcong, prompting a period of introspection. In a letter addressed to the *Breeze*, Maxim engaged in retrospective contemplation, drawing a parallel between his own experiences and the historical voyage of Christopher Columbus. He stated (box 1, folder 18, accession 2147), "When Columbus discovered America, he landed on one of the West Indian Islands. He never reached the mainland." Employing a tone of self-promotion, Maxim boasted about the past allure of the lake, citing the visit of Jerome Bonaparte and his admiration for its beauty. Furthermore, he underscored the current appeal of the region by highlighting the ongoing property improvements being made by Rex Beach.

In asserting his personal significance in shaping the lake's development, Maxim proudly proclaimed, "I myself am one of Lake Hopatcong's discoverers. I first came here twenty years ago this Fourth of July." His perspective portrayed him as an active participant in the discovery and development of Lake Hopatcong, extending beyond the realms of roadways and forestry management. However, it is crucial to note that this viewpoint reflects a colonial perspective, evident in Maxim's reference to Columbus and his disregard for the preexisting history and the migrations and settlements of the region's original inhabitants. As a "discoverer," he oversaw many issues—some trivial and some large.

More missiles in Maxim's home. *Courtesy of the Hagley Museum and Library.*

Upon Maxim's death, *Time* magazine penned a thoughtful obituary, emphasizing, "It is said that Hudson Maxim loved a good fight." Maxim's fervor, whether commendable or otherwise, was undeniably remarkable. The tribute read:

> *As it must to all men, Death came last week to Hudson Maxim, 74, inventor of deadly explosives. It came slowly, quietly—preceded by 24 hours' coma. It found him at his home at Maxim Park, Lake Hopatcong, N.J. It had tried unsuccessfully, many times before, to find him in his laboratory. Although several of his assistants had been blown to bits, he emerged from all his dangerous experiments with only his left hand missing.*

This recognition underscores that Maxim's tenacity was as evident in his commitment to Lake Hopatcong as it was in his work with explosives.

5

THE MODERN PERIOD

In the spring bloom of 2020, as the world plunged into the throes of a global pandemic, our corner of the world felt the change acutely. The familiar rhythms of daily life came to a sudden halt, and we found ourselves enshrouded in the silent hush of a global standstill, an unsettling quiet that was simultaneously peaceful and deeply unnerving. For a lengthy span of months, the world as we knew it seemed to shrink, becoming reduced to the comfortable yet confining walls of our homes, our sanctuaries that served as both a retreat and a prison during this time of uncertainty.

This confinement was broken only by the occasional daring expeditions into the untamed wilderness of supermarket aisles to procure the essentials—from food to the sporadic, prized finds of toilet paper, the gold standard of pandemic-era commodities. As physical interactions dwindled, we found ourselves relying more heavily on technology. It rose to prominence, becoming the backbone of human connection and the pulsating vein of information dissemination. In the wake of this global pause, even traditional activities, such as dining, a beloved pastime at Lake Hopatcong, succumbed to this temporary suspension.

Amid the seemingly ceaseless hum of online chatter, an intriguing call emerged from the Hopatcong Marketplace. The marketplace, a popular venue for local commerce, sought vendors. This announcement struck as both unfamiliar and enticing in the stillness of the pandemic-stricken world. This author's wife, Sandi, an avid artisan, had taken advantage of the imposed retreat to craft an impressive array of artistic pieces. With unceasing

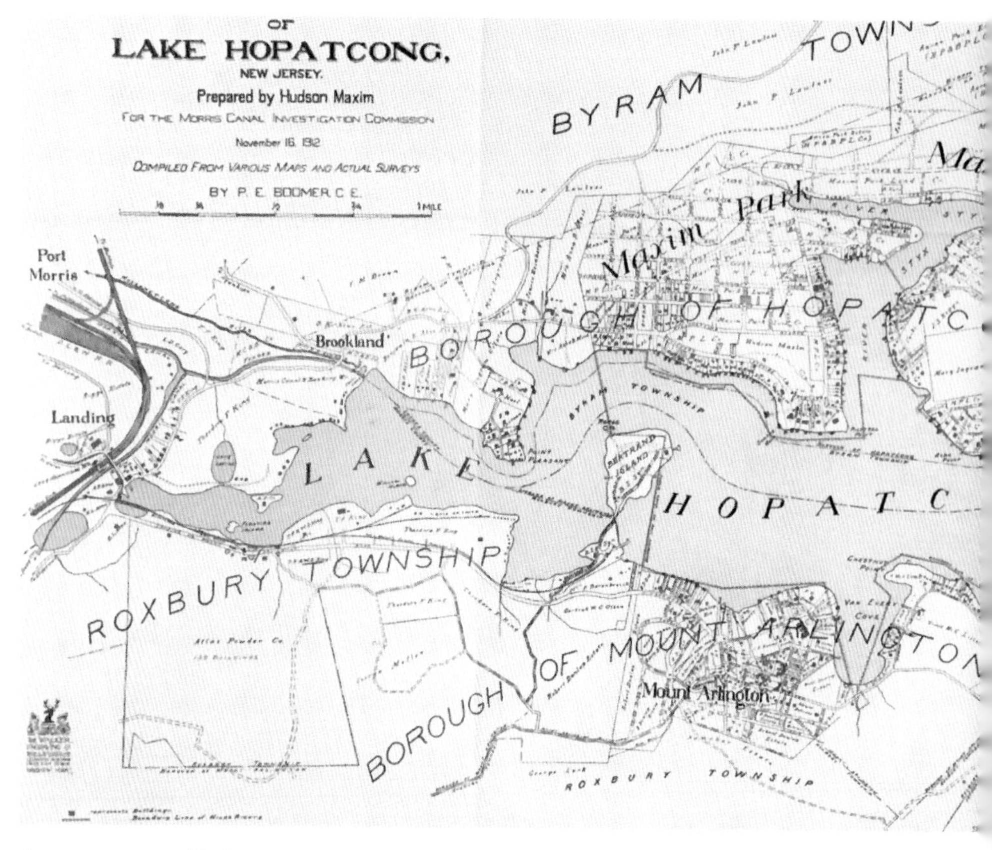

A property map of Lake Hopatcong, New Jersey. *Courtesy of the Hagley Museum and Library.*

passion, she meticulously crafted a collection of awe-inspiring pieces, each distinct and captivating. These artistic masterpieces began to adorn the hidden corners of our home, filling it with a touch of her creativity and transforming our living space into a personal gallery.

The marketplace's call stirred both curiosity and apprehension within us, signifying the first instance in weeks that the world outside our domestic bubble seemed to beckon. Since the onset of the pandemic, our lives had been marked by the blue glow of our computer screens and remote meetings, the rhythm of our newly discovered culinary talents and the adrenaline rush of devising new card games to stave off the encroaching boredom. The prospect of something occurring outside our microcosm, of reconnecting with the broader community, was simultaneously thrilling and nerve-racking.

But nothing could have prepared us for the sheer exhilaration of the first day at the marketplace sometime in May. This author can still feel the

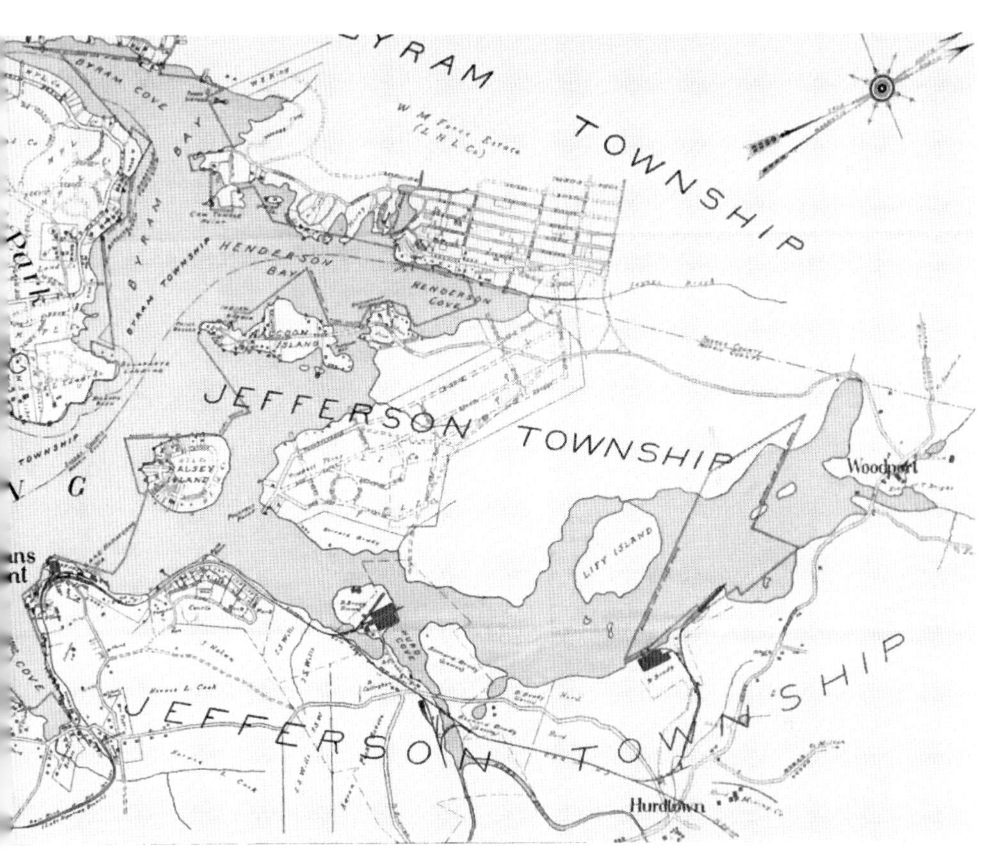

wind that was whipping around us as we set up our tent, maintaining a ten-foot distance from our neighboring vendors—not for lack of space but in deference to the new rules of social distancing. The field was dotted with about ten tents, an expanse of promise under the open sky.

The morning started quietly, almost reverently, but soon, the silent air was punctuated with whispers of conversation and the beginnings of camaraderie. Masked visitors began to drift in, embodying the marketplace's unofficial slogan: "Mosey on by." It was electrifying to witness the community rekindling itself, breaking free from the grip of isolation.

Fast forward a few months, and the Hopatcong Marketplace had transformed into a bustling hub of more than one hundred vendors. Large crowds thronged, reveling in the vibrant commerce, savoring the delicious offerings from food trucks and basking in the simple joy of being part of a thriving community. It felt as if Lake Hopatcong was slowly waking from a long slumber, beginning its journey of healing and rebirth into the future.

Dining at Lake Hopatcong, *Author's collection.*

SwirlsOnline at the Hopatcong Marketplace. *Author's collection.*

Nostalgia, Revival and Enduring Charm

Lake Hopatcong may no longer bask in the same glitz and glamour of the early twentieth century, when grand hotels and luminaries abound on it; however, it still claims its fame. The lake made its big-picture premiere in 1921, with a short "photo-play" that was part of a burlesque series. For film fans, renowned actor Peter Dinklage made *The Station Agent* here in 2003. He filmed numerous segments at Lake Hopatcong, including a notable scene at River Styx Road. In the captivating story, Dinklage delivers a remarkable performance as the lead character, Finbar "Fin" McBride, but the lake community itself is a character. Lake Hopatcong reflects Fin's comfort in solitude and his paradoxical yearning for an accepting community.

For foodies, the lake area has a tantalizing dining scene and a wide array of engaging activities. For instance, among the many establishments hit by this pause was the Windlass, a renowned restaurant situated at Lake Hopatcong. For those who know it, the Windlass is not just a restaurant, it is a cornerstone of the community, a place to gather and share a meal. It is famed for its diverse cuisine, an assortment of dishes designed to suit anything from a quick bite to a five-course meal, complemented by stunning water views that form a captivating visual treat. The menu diversifies from fresh seafood to classic American fare, all equally enjoyable to this amateur food critic. The rustic interior, coupled with the friendly service and an expansive outdoor patio, adds to the charm, creating a welcoming atmosphere. Whether you are going for a romantic dinner, a family outing or a gathering with friends, the Windlass will provide an enjoyable dining experience.

For adventure, besides boating and water activities, Lake Hopatcong is home to a wide array of breathtaking hiking paths. These paths, predominantly composed of gravel and dirt, offer a multipurpose landscape that is perfect for a variety of activities, such as jogging, trekking, taking leisurely walks or cycling. The Lake Hopatcong Trail, for instance, is a straightforward, linear path that travels around twelve miles through Hopatcong without any circular routes. There are several parking areas available for convenient access to different segments of the trail.

In the heart of it is a haven for hiking enthusiasts—Liffy Island, a concealed treasure yearning to be found. For those who crave genuine outdoor escapades, this hiking destination will impress you. To embark on this enchanting journey, venture forth to the James Leach Boardwalk Trail at Prospect Point, which

This page, top: River Styx Bridge at Lake Hopatcong featured in a scene of *The Station Agent*. *Author's collection.*

This page, bottom: The Windlass Restaurant at Lake Hopatcong. *Author's collection.*

Opposite: Lake Hopatcong trail. *Author's collection.*

is the threshold to this wilderness retreat. This trail seamlessly merges with the sprawling Lake Hopatcong Trail network, promising a panorama of experiences tailored to the outdoor enthusiast's soul.

The James Leach Boardwalk Trail, a sinuous path nearly three miles in length, will lead you through carefully maintained trails. As your journey through this scenic landscape unfolds, you'll find yourself cradled in the calm of nature's grip. The trail wends its way through lush surroundings. After a roughly one-mile ascent, a charming bridge emerges, a portal to Liffy Island. Traverse this bridge, and you'll step into a realm of natural marvels.

The boardwalk-inspired bridge offers captivating views of the lake. While crossing, visitors can participate in birdwatching, observe aquatic life and marvel at the impressive landscape. Upon reaching the island, visitors can find numerous unmarked trails that are available for exploration. Adventurous visitors can wander off these paths to get closer to the water. Despite the lack of marked trails, the island is easy to navigate. The total distance of the hike is slightly under three miles. It is an unforgettable odyssey.

While there are no camping spots directly on Lake Hopatcong, there are wonderful campgrounds nearby. Located in the northern region of New Jersey, close to Andover and nine miles from Lake Hopatcong, is the Panther Lake Camping Resort, a great place for those who love the outdoors. This expansive resort is situated right at the southern tip of Panther Lake, which spans forty-five acres, creating a unique and immersive camping environment. The resort has room to house as many as 410 vehicles. However, it's not just limited to RVs, as there are spaces dedicated specifically for those who prefer tent camping. The resort also has a wide array of land activities for entertainment. There's a playground for kids to enjoy, and adults have numerous games to choose from, such as tennis, basketball, volleyball and shuffleboard. All in all, the Panther Lake Camping Resort provides an exciting and versatile outdoor experience for all ages, and most importantly, it is near Lake Hopatcong.

Hopatcong State Park is a versatile outdoor spot that's enjoyable throughout the year. During the summer, it comes alive with a variety of activities. The park provides plenty of picnic tables and grills, making it great for gatherings. There are a range of athletic activities like beach volleyball, basketball, a playground for the kids and places to swim. Hopatcong State Park offers something for everyone during the vibrant summer season. During the winter, people gather to enjoy the frozen scenery.

Hopatcong State Park also permits fishing under specific guidelines. Night fishing is allowed from March 15 until the Friday before Memorial Day, running from sunset to midnight. Between March 15 and the second Friday in May, night fishing can be enjoyed on the sandy beach area. However, from the subsequent Saturday until the Friday preceding Memorial Day, this sandy area is off-limits for fishing. In this period, night fishing is limited to areas beyond the rock jetties. Anyone between the ages

Right: A Lake Hopatcong trail. *Author's collection.*

Below: Hopatcong State Park. *Author's collection.*

Hopatcong State Park rules. *Author's collection.*

of sixteen and sixty-nine who wishes to fish is required to have a valid New Jersey fishing license. Additionally, the park facilitates boating with the provision of a launch ramp.

The lake also hosts various annual events that allow the community to come together. Of note, the Hopatcong Marketplace has now established itself as an annual tradition, transforming several weekends into bustling celebrations of local trade and craft. Equally anticipated is the Hopatcong Days at Modick Park, an annual carnival overflowing with a smorgasbord of food, an array of vendors and exhilarating rides. Every year, the shores of Lake Hopatcong come alive during the Lake Hopatcong Block Party. Hosted by the Lake Hopatcong Foundation, this grand event stitches together community organizations, local merchants and vendors, curating a vibrant and engaging public fair. As of 2023, visitors are treated to a range of entertaining activities, such as exploring animal exhibits that feature alpacas and other creatures. A variety of kid-friendly activities are usually planned, including fun crafts, exciting scavenger hunts and large-scale games.

The Lake Hopatcong Yacht Club, with its rich history and community, has become an integral part of the regional landscape. Often serving as the hub for local events and gatherings, its influence stretches beyond just the yachting enthusiasts. The origins of the club date back to 1907, when the construction of a clubhouse on Bertrand Island began. Like any ambitious project, this journey was not without its trials. The construction faced numerous obstacles, including unexpected delays and cost escalations. Initial budget estimates were swiftly surpassed, with the final construction cost soaring to somewhere between $7,000 and $7,500. To offset these burgeoning expenses, the club resorted to implementing initiation fees and annual dues. This not only raised necessary funds but also instilled a sense of shared responsibility and community spirit among the members. Demonstrating their dedication to the club, members went to great lengths to contribute, even donating timber for the clubhouse construction.

But the clubhouse was not just about the timber. Additional expenses, such as stone for the foundation, plaster for the walls, a dock for the boats and a flagpole to symbolize the club's presence, were also taken into account. These elements, though seemingly minute, were indispensable in shaping the clubhouse's identity. By the time everything was accounted for, the total expenditure on the clubhouse project came to $9,860. Ever since its establishment, the Lake Hopatcong Yacht Club and its clubhouse have served

Hopatcong Days at Modick Park. *Author's collection.*

as enduring symbols of the community's persistent spirit. They have been the foundational pillars of the region, promoting social interaction, water activities and a collective appreciation for the splendid Lake Hopatcong.

The clubhouse was officially inaugurated on July 9, 1910. It boasted ample room space, a versatile ballroom featuring a movable stage, expansive porches and its own power supply. The Ladies Auxiliary tastefully furnished the interior, and club members gifted various items, including two sizable anchors, adding personal touches to the space. From its very first day, the newly constructed clubhouse was a hive of activity. The inaugural day saw four hundred club members enthusiastically meandering down the singular dock, ready to compete in the weekend's boat races. The celebration continued with delicious meals savored on the newly decked out wrap-around porch.

The clubhouse swiftly transformed into a lively center for all lake-related activities, solidifying the yacht club's significance in the community. This marked a fresh chapter in the club's history, setting the stage for its future growth and prosperity. Notable personalities like Hudson Maxim and Rex Beach were once part of this esteemed membership. Even today, more than one hundred years later, the club continues to stand strong. Operating on a seasonal basis, the Lake Hopatcong Yacht Club opens its doors to members from May to September each year, upholding its cherished tradition.

Even for those who reside nearby and have no direct connection with the Yacht Club, it's difficult to ignore its existence, particularly when it illuminates the sky with a spectacular fireworks display every Fourth of July. Stepping onto the lake isn't a prerequisite to bask in its splendid display. Lake Hopatcong, with its rich surroundings of various trees, offers a captivating spectacle for residents right from the comfort of their backyards. The dazzling play of light against the abundant foliage transforms the nearby spaces into enchanting realms, filled with an arresting array of verdant greens and fiery reds. It's a spectacle that seemingly paints the landscape anew each time, imbuing the familiar scenery with a magical aura during each unique show.

The lake itself boasts an array of enticing attractions on the water. Various types of otium cruises grant passengers awe-inspiring, panoramic views that highlight the lake's innate grandness. Some cruise services combine this stupendous scenery with food and music, allowing passengers a unique opportunity to enjoy a meal amid the picturesque majesty. Event cruises cater to significant occasions, such as weddings, anniversaries or corporate events. They offer an upscale and intimate ambiance for festive gatherings.

Sightseeing cruises offer a leisurely experience to appreciate the lake's historical sites and natural landmarks. Furthermore, these cruises offer a

Fireworks over Lake Hopatcong. *Author's collection.*

unique opportunity to explore and understand the delicate ecosystem of Lake Hopatcong. The guides provide informative commentary about the diverse flora and fauna that call the lake their home. Passengers gain an understanding of the interconnectedness of the lake's ecosystem and its importance for the local environment. These journeys serve as both relaxing tours and educational expeditions.

For Lake Hopatcong residents, there are exclusive opportunities to experience the tranquil essence of the lake. Aside from Hopatcong State Park, the local municipality doesn't provide any publicly open beaches. Instead, various homeowner organizations manage the smaller beaches surrounding the lake. Places like Crescent Cove and Capp Beach offer private sections where residents can bask in the comfort of their exclusive lakeside retreat for an annual fee. Beyond offering a peaceful escape, these beaches host regular events such as movie nights and game nights, fostering a lively and engaging atmosphere that perfectly captures the joyful essence of life at Lake Hopatcong.

On the shoreline, there is a lively mini-golf course in Jefferson Township that summons people of all ages. This eighteen-hole jewel is deeply rooted

A replica cruise boat at Lake Hopatcong, *Author's collection*.

Crescent Cove at Hopatcong. *Author's collection*.

Mini golf course in Jefferson, New Jersey. *Author's collection.*

in the aesthetic spirit of its stunning surroundings, providing a sweet merger of a playful challenge, serene relaxation and delight. It is a welcoming refuge for families, groups of friends and individuals, as the easygoing rhythm of mini-golf melds seamlessly with the grand design. As ethos meets pathos, please excuse my indulgence as I wax poetically about the place where I proposed to my now wife on a gorgeous summer day.

Just to paint a picture: imagine that with a crisp lake breeze tousling your hair, you get to perfect your swing while taking in the breathtaking views of serene waves, gracefully drifting boats and friendly passersby, all the while becoming engrossed in the spectacular panorama. The ambiance is further sweetened by nature's own melody, with ducks and other birds contributing a pleasing, natural chorus to your moment of leisure. The mini-golf course transcends its role as a simple recreational activity, emerging as a symbiotic blend of sporting fun and soothing lakeside absorption. It serves as a medium through which memories are artfully shaped. It places emphasis not just on the game but also on the embracing, immersive experience amid nature's gentle cradle.

For older residents, Lake Hopatcong still holds memories of its past family-oriented attractions. Bertrand Park, a popular amusement park located on a

Lake Hopatcong Golf Club. *Author's collection.*

small peninsula in the lake, opened in 1910 and gradually introduced rides. In the 1920s, Bertrand Island Amusement Park experienced significant growth, offering an array of thrilling attractions. These included favorites, like the Ferris wheel, roller coaster, Dodgems (bumper cars), aeroplane swing, the Old Mill (later renamed the Lost River), sightseeing boats, the Whip and more. The roller coaster, built in 1925, was notably the first of its kind in Northern New Jersey. The addition of the famous Illions Supreme carousel in 1937 further enhanced the park. Over the years, new rides and games were continually added to keep up with trends and preferences. It gained popularity and even earned the nickname "Little Coney."

However, as Lake Hopatcong became less of a tourist destination, Bertrand Island Amusement Park lost its appeal. Its final owner, Gaby Warshawsky, ultimately decided to sell the property. It was later subdivided and transformed into residential land. The park officially closed its doors in September 1983, marking the end of an era. Today, a row of townhouses stands in its place in Mount Arlington. The park's remains can be seen as the backdrop for Woody Allen's film *Purple Rose of Cairo*. Now, only a stone marker near a community pool house remains. Although it no longer exists, for those who had the chance to experience the park, there remains a lasting sense of nostalgia and cherished memories.

The Interdependence of Lake Hopatcong's Ecosystem and Community

There is no way around it, human actions exert significant influence on lakes. It is similar to a garden, where humans can nurture and protect it in order to enjoy the beauty and bounty it provides. If the garden is neglected, there is a mess to clean up. By gaining insight into the profound significance of human engagement, we deepen our understanding of the duty to protect it. Recognizing the intricate connections between individuals, communities and nature can lead to a collective effort to ensure the sustainability of this ecosystem and the neighboring communities it supports. Ultimately, this understanding can foster a sense of stewardship and responsibility. As Neil deGrasse Tyson put it: "We are all connected; To each other, biologically. To the earth, chemically. To the rest of the universe atomically." So, basically, we're all (including humans and nonhumans) part of the same family.

By understanding the interconnectedness between humans, nature and communities, we can develop strategies that are designed to create a more sustainable future. This includes conserving resources, reducing waste and protecting natural habitats. Additionally, we can create incentives for more sustainable practices and invest in renewable energy sources. This effort also involves responsible recreational activities, prudent resource management and the cultivation of stewardship. It necessitates educating and engaging the public in the importance of a healthy Lake Hopatcong ecosystem. Later in this chapter, I highlight the amazing organizations that have taken up this cause. Like any environment, it calls for a partnership between people and nature.

A study called "Evaluating the Effects of Historical Land Cover Change on Summertime Weather in New Jersey" found: "Human activity, particularly over the last few centuries, has left a profound footprint on the landscape." It notes the following human activity: "We have cut down native forests and replaced them with agriculture and pasture land to feed us and our livestock. We have seriously degraded the wetland ecology that once protected our local communities from flooding and filtered dangerous pollutants from our drinking water." The study underscores the importance of recognizing the profound footprint left by human activity on places. For instance, even the construction of dams and reservoirs can cause the displacement of many species of animals and plants as well as the destruction of valuable habitats. It calls for an awareness of the consequences actions can have.

Regarding its specific relevance to Lake Hopatcong, there has been a notable surge in development around the lake, resulting in the construction of new homes and businesses. This heightened activity has placed additional stress on the local environment, as the lake's resources and infrastructure face increased demands. Zooming out to a broader perspective, New Jersey has witnessed a temperature rise of approximately one to two degrees Fahrenheit over the past century. This aligns with, and in some instances, exceeds the global warming trend. The primary driver of this change is human-induced climate change, which has manifested in the form of more frequent and intense heatwaves.

Heatwaves are now occurring over six times per year, on average, throughout the United States. In 2023, New Jersey broke some of its own records for the hottest days on record. Globally, September 2023 ranks among the hottest months ever documented. Such extreme weather events pose significant health risks, particularly for vulnerable populations such as the elderly and those with preexisting health conditions. A warmer winter has several practical nuisances, including the potential for tick populations to remain active for more extended periods than usual.

Alongside rising temperatures, New Jersey is also grappling with the impacts of climate change in the form of rising sea levels. As the Earth's temperature rises, ice caps and glaciers melt, leading to the expansion of oceans. Coastal areas like those in New Jersey are now facing higher sea levels, increasing the threat of flooding and erosion. Furthermore, climate change has caused shifts in precipitation patterns in New Jersey. The timing, intensity and distribution of rainfall have been altered. Some regions have experienced heavier downpours, increasing the risk of flash floods and water-related damage. Conversely, other areas may endure extended periods of drought, impacting agriculture, water resources and overall ecosystem health.

The implications of global warming extend even further, impacting a wide spectrum of living organisms. Take, for example, lily pads, which, along with countless other plant and animal species, can be profoundly affected by climatic fluctuations linked to global warming, such as shifts in temperature and precipitation. An increase in temperature can extend the lily pads' growth period. If the conditions are conducive, this could lead to an increase in their numbers and more of them covering the water's surface.

Lily pads on Lake Hopatcong. *Author's collection.*

However, a potential drawback of increased temperatures is accelerated evaporation, which could result in lower water levels in the habitats of lily pads, negatively impacting their survival. Furthermore, the changes in the timing and intensity of certain weather events—earlier onset of spring, for instance—can disrupt the typical life cycle of lily pads. Such disruptions could influence their reproductive processes and overall lifespan. This is just one example of how global warming can have different effects on different kinds of plants and animals.

TRANSFORMATIONS, HARDSHIPS AND SOCIOECONOMIC SHIFTS

Following the passing of Hudson Maxim in 1927, his legacy remained, but his beloved region underwent significant transformations. Like the rest of New Jersey, Lake Hopatcong experienced the hardships of the 1930s. During this time, economic conditions were dire, with a substantial portion of the state's population relying on New Deal programs for survival. However, government funds began to dwindle, casting a dark cloud over New Jersey, including Lake Hopatcong, amid notable events, such as the Lindbergh kidnapping, the Hindenburg disaster and the "War of the Worlds" radio broadcast.

The challenges persisted into the 1940s, with World War II bringing further difficulties. Gasoline rationing curtailed travel, and limited supplies made it challenging for hotels to remain open. The few surviving hotels gradually closed their doors, culminating in the destruction of the last operating hotel by fire in 1972. Consequently, valuable properties fell into disrepair, and even Maxim's main house met a similar fate in the 1950s.

Amid these hardships, the social fabric of the lake community changed. With the absence of hotels, Lake Hopatcong evolved into a year-round residential community, prioritizing practical housing over extravagant accommodations. In the borough of Hopatcong, the year-round population has continuously increased. Starting with a mere 75 inhabitants in 1900, the population rose to 146 in 1910; 179 in 1920; 534 by 1930; 660 in 1940; 1,172 in 1950; 3,391 in 1960; 9,052 in 1970; 15,531 in 1980; 15,666 in 1990; and ended at 15,888 by the turn of the millennium in 2000. This shift in identity, combining remnants of the past with present needs, brought stability to the lake community. It transformed the perception of the lake from a mere escape from urban life to a place where individuals could establish their homes and forge new identities.

The issues of development and pollution have posed continuous problems. Dialogues concerning zoning regulations started as early as the 1920s in an attempt to control expansion and safeguard the lake's ecosystem. Untreated sewage, coupled with agricultural and industrial waste, contributed significantly

Educational materials at Hopatcong Days by the Hopatcong Environmental Commission. *Author's collection.*

to water pollution, escalating it to a critical issue. The rise in commercial and residential development compounded the problems, leading to the destruction of natural habitats, significant soil erosion and extensive deforestation.

The advent of zoning discussions in the 1920s marked the initial attempts to find a balance between development and preservation. The goal was to effectively manage growth around the lake while ensuring the conservation of its natural beauty and ecological balance. The initiatives involved comprehensive land-use planning, with proposals aiming to mitigate the adverse impacts of unplanned development.

The 1980s saw the consequences of these changes, as detailed in a 1984 *New York Times* article. In the article "The Environment," Leo H. Carney opines, "One of the major environmental and economic challenges of this decade is said to be the preservation of Lake Hopatcong, New Jersey's largest semi-enclosed body of fresh water, which has been succumbing to the ravages of overdevelopment." Carney highlights some of the conditions that the community was facing and would need to deal with.

There have been a number of harmful elements introduced into Lake Hopatcong over the last century. They have come from agricultural runoffs, septic systems and lawn fertilizers. It could be worse. Some of the most common harmful elements found in other lakes are lead, mercury, perchlorate, radium, selenium, silver and uranium. These elements brought by anthropogenic modifications can be devastating. Other, more polluted lakes underline the necessity of maintaining a delicate balance at Lake Hopatcong. With finite resources, the urgency to retain the lake's natural state becomes more crucial.

Lake Hopatcong now grapples with eutrophication, a natural but problematic aging process of water bodies. *Eutrophic* is derived from a Greek word that translates to "well-nourished," signifying a series of physical, chemical and biological transformations in a freshwater body. Eutrophication is primarily caused by agricultural runoff, such as fertilizers, wastewater discharge from houses, gas emissions and even pet waste. The effects manifest through multiple signs, including silting, delta formation, reduced oxygen, increased water turbidity and unchecked growth of algae and aquatic weeds. Eutrophication, which can span decades or centuries, marks the gradual transformation of a lake or pond into a swamp, marsh, bog and, ultimately, a meadow.

Early indicators of eutrophication include silting, or the accumulation of fine sediments, which diminishes water depth and alters aquatic life. Similarly, delta formation—sediment deposition where a stream meets the lake—can increase shallowness and spur terrestrial vegetation growth. Decreased oxygen levels in the water, largely caused by excessive algae, can threaten the survival of fish and other oxygen-dependent aquatic life. Eutrophication can also cause water turbidity, limiting light penetration essential for submerged plants and impairing visibility for aquatic creatures.

Furthermore, rapid growth of aquatic weeds can overrun shorelines and disrupt recreational activities. If unchecked, these conditions can lead the water body to become a less hospitable environment, gradually transitioning into a swamp or marsh, then a bog and, finally, a meadow. To preserve Lake Hopatcong's ecosystem and its recreational value, it's vital to proactively manage and mitigate the effects of eutrophication.

Reports from the Lake Hopatcong Regional Planning Board spanning several decades provide a deeper understanding of the ecological dynamics of Lake Hopatcong, acknowledging that lake water quality and levels can be significantly influenced by both natural and human activities, including storm runoff and intense recreational use. Notably, the lake's hydrology is crucial due to its limited outflow through the Musconetcong River, which renders it more susceptible to the effects of increased human activity, as restricted outflow hinders the natural turnover of water. This understanding of the lake's hydrological dynamics is particularly pertinent given the documented warming temperatures in the two counties encompassing Lake Hopatcong. In fact, historical temperature data indicates a notable increase in average winter temperatures over the past century, with Sussex County experiencing a rise of 2.6 degrees Celsius (4.7 degrees Fahrenheit) since the winter of 1895–96 and Morris County observing a slightly higher increase of 2.7 degrees Celsius (4.9 degrees Fahrenheit) in the same period. Such scientific investigations serve to quantify the consequences of human activities on the lake's ecological state.

To battle eutrophication, the Lake Hopatcong community started implementing mitigation measures, such as storm management systems. Conservation efforts in the 1990s and zoning regulations began to alleviate environmental issues, like water pollution and habitat loss, demonstrating an ongoing commitment to the lake's preservation. These initiatives included addressing pollution to ensure Lake Hopatcong's long-term ecological health. Yet many regions around the lake still depend on septic systems due to construction challenges posed by the rocky terrain.

Nevertheless, the towns around Lake Hopatcong, such as Hopatcong and Jefferson, now have guidelines for septic systems. When a dwelling is modified to include more bedrooms or a commercial or nonresidential building's use or size changes in a way that it puts more strain on a septic system, that system must be updated according to current codes. The applicant is accountable for demonstrating the number of bedrooms in both the previous and proposed residences and the septic system's previously approved capacity.

A certificate of compliance, based on a licensed professional's written statement, is required before using new or altered septic systems to ensure they meet permit conditions and New Jersey laws. Septic systems must now be fully installed and authorized by the health officer on lots smaller than fifteen thousand square feet (about half the area of a large mansion) prior to a footing inspection. To construct, reconstruct or extend septic tanks or other individual sewage disposal systems, a yearly renewable license from the New Jersey Board of Health or its representative is required, which can be revoked for noncompliance.

These measures provide a proactive approach to maintaining the cleanliness and quality of the lake's water. By focusing on septic systems that are leaky, outdated or overtaxed, these measures aim to prevent harmful substances from entering the lake and causing pollution. Faulty septic systems can be hazardous, as they could potentially release untreated sewage into the environment, including the lake. Untreated sewage contains bacteria, viruses and other pathogens that can cause serious health problems—not to mention harm to the aquatic ecosystem.

Aside from updated septic systems, the landscape surrounding Lake Hopatcong has experienced significant transformations in recent years. Just in 2019, an article on NJ.com titled "Lake Hopatcong's Disappearing Boathouses Take Regional History with Them" reported:

> *But like the gilded hotels and mansions that succumbed to fire, neglect, and development over the years, the boathouses of Lake Hopatcong are in danger of fading into history, and with them, local officials fear, will go some of the region's rich past.*

These sentiments highlighted the supposed loss of the lake's twentieth-century character, but this was about to change just a year later.

The year 2020 brought about yet another transformation within the lake community, largely influenced by the COVID-19 pandemic and people looking for open spaces. Rising appreciation for natural environments came with increased residential interest. Many boathouses have acquired new owners and are undergoing refurbishment. Capuzzo explains how "the lake is now enjoying a second act, as more people have discovered its vacation-like charms during the pandemic." The COVID-19 pandemic, coupled with these nostalgic narratives, has once again reshaped the social fabric surrounding Lake Hopatcong. The evolving ownership and revitalization of boathouses reflect the ongoing process of adaptation and rejuvenation within the lake community.

To augment the impact of visitors, Lake Hopatcong could benefit from the utilization of more open spaces, akin to the approach of the Tennessee Valley Authority as discussed by Frank E. Smith in *Land Between the Lakes*. Smith writes about the use of open areas developed by the Tennessee Valley Authority in Kentucky and Tennessee. He discusses the utilization of such spaces by the Tennessee Valley Authority to prevent excess pollution. In the case of Lake Hopatcong, access to the lakefront is predominantly controlled by private owners, while the public can access the lake through Hopatcong State Park, small membership beaches and private businesses. Maxim's warning about the lake's susceptibility to pollution due to its irregular shape

Lake Hopatcong homes. *Author's collection.*

and the presence of homes directly encircling the shores is noteworthy, as mentioned in chapter 4 of this book.

Keeping a lake healthy is more than just a local environmental concern; it is one with global implications. Lakes play a significant role in global warming, despite covering only a small fraction of the Earth's surface. Research conducted by the Jefferson Project at Lake George reveals that lakes have a greater impact on global warming than oceans, primarily due to their ability to sequester carbon. In fact, lakes have a greater ability to store carbon than all the oceans combined. While larger lakes often serve as popular vacation destinations, human activities and interactions with these bodies of water can have profound effects on their ecological health, as well as the broader climate system.

Drawing a comparative analysis between Lake George, a relatively large body of water (28.5 square miles), and the smaller Lake Hopatcong (4 square miles) provides insights into the parallel challenges posed by global warming and human interaction. For example, both lakes have experienced a change in water depth. Causes vary by region and include unsustainable water consumption, shoreline construction, changes in rainfall, temperature fluctuations and more. Luckily, the Lake Hopatcong Commission actively supports ways to preserve efforts and implements regulatory measures to shield the lake's integrity and water. This includes restrictions on motorboat speed and no wake zones, which help reduce erosion and sedimentation. It means that if Maxim was still around, he would need to be more responsible with his motorboat.

NEW TOOLS AND ORGANIZATIONS FOR A NEW ERA

The twenty-first century has seen a series of contemporary initiatives aimed at preserving Lake Hopatcong. Some of these measures are physically apparent. For instance, every five years, as stipulated in the 2011 management plan, the New Jersey Department of Environmental Protection (NJDEP) strategically decreases the water level of Lake Hopatcong by five feet. This deliberate action not only ensures the maintenance and periodic inspection of vital shoreline infrastructure but also plays a crucial role in the upkeep of the dam located at Hopatcong State Park. Such proactive measures are modern examples of forward-thinking strategies employed to ensure the long-term health and preservation of the lake. By drawing on these institutional developments and conservation efforts, a new multifaceted approach to safeguarding Lake Hopatcong's environmental integrity has been created.

The establishment of the Lake Hopatcong Commission in 2001 marked a pivotal moment in the ongoing conservation efforts for New Jersey's largest lake. The Lake Hopatcong Protection Act created the commission and tasked it with improving Lake Hopatcong and its surrounding area. It is governed by an eleven-person volunteer board. According to the Lake Hopatcong Commission's own website, the group is dedicated to preserving the lake's water quality and protecting its natural, scenic, historical and recreational value. Principal duties include conducting water quality research, devising an extensive lake management strategy, coordinating with diverse government bodies and initiating public awareness campaigns about the importance of conserving Lake Hopatcong's resources. This commission brought a notable ally to Lake Hopatcong.

The positive outcomes resulting from the establishment of the commission are evident in the various accomplishments and advancements made in the conservation and management of Lake Hopatcong. For instance, the Environmental Protection Agency (EPA) granted the commission federal funding of $745,000 in 2005. This financial support enabled the implementation of targeted initiatives aimed at enhancing water quality and addressing the detrimental effects of phosphorus pollution. The commission's proactive approach to lake management, coupled with its integration of innovative strategies, stands as a vital component in shaping the future of Lake Hopatcong. Through their dedicated efforts, the commission has implemented measures to prevent and mitigate water pollution, address invasive species concerns, promote sustainable recreational activities and preserve the lake's historical landmarks. These initiatives have helped maintain the ecological balance of the lake, enhance its recreational value and ensure its long-term vitality.

Additionally, the formation of the Lake Hopatcong Foundation in 2013 further stresses the commitment to preserve, protect and enhance the lake's natural resources. According to the foundation's website, "The Lake Hopatcong Foundation works to foster a vibrant and healthy Lake Hopatcong and its surrounding community. We do this through a variety of programs and initiatives in the following areas: environment, education, community and historical preservation, public safety, recreation, and arts and culture." Alongside promoting safe boating practices, the foundation has also undertaken various initiatives designed to improve the lake's water quality and overall ecological health.

The Lake Hopatcong Foundation also helps restore the character of the region. In 2014, it purchased and renovated the historic Lake Hopatcong train station. This acquisition marked a new chapter for the foundation and the region, preserving the building and providing a suitable location for future educational and cultural programs. Following the purchase, the foundation received an overwhelming response from the community, with approximately fifty volunteers stepping forward to assist in clearing the grounds surrounding the building. The goal was to restore the station to its original 1911 appearance, uncovering original details like plaster walls, fieldstone and the ticket window area. With community support, volunteers helped clear the grounds, demonstrating the community's willingness to contribute to restoration efforts.

By drawing on these institutional developments and conservation efforts, a multifaceted approach to safeguarding Lake Hopatcong's environmental integrity has been established. The confluence of scientific research, institutional frameworks and grassroots initiatives has engendered a paradigm shift in the perception and stewardship of Lake Hopatcong. Through these concerted efforts, the lake's natural, historical and leisurely values are being

The historic Lake Hopatcong Train Station. *Author's collection.*

guarded while promoting a harmonious coexistence between human activities and the lake's ecological systems. However, ongoing vigilance, continued research and adaptive management practices are essential to effectively address emerging challenges and sustain the environmental sustainability of Lake Hopatcong in the face of evolving societal and ecological dynamics.

In 2019, Lake Hopatcong experienced a significant outbreak of harmful algae blooms, which enforced shutdowns in the area. The outbreak of algae was caused by a combination of specific factors. First, above-average rainfall during the spring season increased water volume and caused nutrient runoff into the lake. Subsequently, intense storms followed by periods of calmness and ample sunlight characterized the summer season. This combination of weather conditions created an ideal environment for the rapid growth and proliferation of cyanobacteria, resulting in the significant outbreak of harmful algae blooms in Lake Hopatcong.

In response, the Department of Environmental Protection (DEP) recommended the closure of public beaches affected by the advisory. The public was urged to avoid activities involving direct contact with the water, such as swimming and water-skiing. The consumption of water or fish caught in the area was also discouraged due to the potential health risks. People were advised to exercise caution, even when participating in secondary contact activities.

Health warnings were issued regarding the potential effects of ingesting cyanobacteria. These include a variety of symptoms in humans, such as headaches, sore throat, stomach pain, nausea, a dry cough, diarrhea and blisters around the mouth. Direct skin contact with the harmful algae could also lead to a rash. In addition, animals exposed to cyanobacteria could exhibit signs of lethargy, stumbling and a loss of appetite.

To effectively address the challenge of HABs and safeguard the ecological integrity of Lake Hopatcong, a multifaceted approach was required. This entailed implementing robust nutrient management strategies to mitigate excessive phosphorus and nitrogen inputs into the lake, such as reducing agricultural runoff, managing stormwater and promoting responsible fertilizer use. Additionally, ongoing monitoring programs are crucial to track water quality indicators, detect the early signs of algal blooms and inform timely management interventions. Public education and outreach initiatives play a vital role in raising awareness among lake users and residents.

In order to address the challenges posed by algae growth and weed proliferation in Lake Hopatcong, the Lake Hopatcong Foundation has enlisted

Lake Hopatcong closed. *Author's collection.*

Recreation at Hopatcong State Park. *Author's collection.*

the expertise of Princeton Hydro. This collaboration aims to implement mechanical weed harvesting, which has been identified as the most cost-effective and environmentally sustainable approach to managing nuisance weed densities in the lake. With the oversight of the New Jersey Department of Environmental Protection (NJDEP), this joint effort is focused on maintaining the ecological balance of the lake and preserving its recreational value.

Princeton Hydro plays a vital role in the welfare of Lake Hopatcong. However, due to budgetary constraints in 2009, there were significant reductions in plant biomass and phosphorus removal, resulting in the unfortunate layoff of the commission's full-time operation staff. Nonetheless, in 2020, Princeton Hydro initiated a large-scale application of Phoslock, a clay-based technology that effectively neutralizes nutrients. This notable treatment has been recognized as the largest Phoslock application in the northeastern region of the United States.

By utilizing mechanical weed harvesting, the collaborative efforts between the Lake Hopatcong Foundation and Princeton Hydro aim to control the density of nuisance weeds, mitigating their impact on the lake's ecosystem. This method involves the use of specialized equipment to physically remove excessive weed growth, promoting a healthier aquatic environment for both flora and fauna. Furthermore, the implementation of Phoslock technology demonstrates Princeton Hydro's efforts to improve water quality in Lake Hopatcong. Phoslock acts as a natural filter, binding and immobilizing phosphorus in the lake, thus reducing the nutrient load that fuels excessive algae growth.

The collaboration between the Lake Hopatcong Foundation and Princeton Hydro, under the oversight of the NJDEP, exemplifies a proactive approach to tackling the challenges that Lake Hopatcong faces. Through the implementation of environmentally responsible methods, such as mechanical weed harvesting, providing educational opportunities for the public and the utilization of innovative technologies like Phoslock, the goal is to foster a thriving and harmonious ecosystem within the lake.

Nevertheless, the problem of algae persists, underscoring the necessity for ongoing proactive initiatives in the future. Just in 2022, water tests in Crescent Cove again confirmed the presence of these harmful algae blooms, triggering an "advisory." Subsequent testing identified samples that, for the first time that season, exceeded the milder "watch" threshold applied to other parts of the lake. The DEP issues an advisory when samples show concentrations of at least 80,000 cells per milliliter of cyanobacteria, the algae responsible for the blooms. The Crescent Cove sample displayed a concentration more than double this limit, reaching 171,500 cells per milliliter.

Fortunately, in late 2022, there was a positive development with the announcement of nearly $10 million in federal COVID-19 relief funds allocated for improving water quality in New Jersey. A substantial portion, over $1.75 million, was earmarked specifically for combatting harmful algae blooms in Lake Hopatcong. The distribution of this aid, totaling $9.95 million, was managed by the DEP and allocated to twenty-three entities within the state.

Notably, the Lake Hopatcong Commission is set to receive $1 million, while Mount Arlington and the Morris County Park Commission are each allotted approximately $750,000 to address pollution within the lake's watershed. These funds, originating from the American Rescue Plan Act of 2021, will be directed toward enhancing water quality and mitigating the adverse effects of stormwater runoff on public lakes throughout the state.

Above: The Floating Classroom on Lake Hopatcong. *Author's collection.*

Left: A Lake Hopatcong commission boat. *Author's collection.*

Although Lake Hopatcong may not be swimming in money, these funds will undoubtedly go a long way in addressing its water quality concerns.

In 2018, the Lake Hopatcong Foundation launched the Floating Classroom as a means to raise awareness and provide educational opportunities. This educational platform takes place aboard the *Study Hull*, a purpose-built forty-foot vessel specially designed to facilitate learning about the environment and marine ecology. The Floating Classroom's core experience lies in its interactive ninety-minute summer cruises. During these enlightening voyages, participants, both children and adults, dive into hands-on learning activities. They conduct intriguing experiments that include testing the water clarity of Lake Hopatcong with secchi disks and studying the intricate world of plankton collected from samples drawn from the lake's water. For instance, one outing included a tour of the lake, learning about the different species of fish and a discussion of the importance of wetland ecosystems. These excursions, rich in knowledge and environmental awareness, set sail every Monday from the Hopatcong State Park. The schedule generally spans from July 11 to August 29.

Just for Fun: The Mythical Threats

This section explores the significance of local folklore centered on a mythical creature believed to have inhabited the region. From the earliest inhabitants, such as the Lenape tribe, to the arrival of settlers, stories have been passed down through generations, describing a formidable and powerful being.

According to "Monsters of New Jersey: Mysterious Creatures in the Garden State," the Lenape spoke of a monster with "a gigantic horse-like head, antlers, and horns." Legend has it that this creature died by falling through the ice. Later, when new settlers arrived, they embarked on a quest to locate its body, and as the legend goes, they succeeded in doing so. However, intriguingly, despite the legend having a conclusion with the apparent "recovery" of the monster's body, people have persisted in sharing accounts of being seized and pulled underwater by this mysterious creature. I guess you could say that its spirit lives on.

These persistent reports of lake monsters, spanning both the past and the present, allow humans to attribute agency and instill a sense of fear into nature that transcends the boundaries of scientific understanding. Wherever nature exists, narratives concerning its ominous aspects can be found, serving as a testament to humanity's inclination to impart intrigue and danger to the natural world. And hey, we get a few good stories out of it.

Considering that a struggle exists to understand nature, there is an inclination to give it agency. In *Beyond Nature Writing: Expanding the Boundaries of Ecocriticism*, Karla Armbruster states, "Landscapes, unlike monsters, are devoid of agency; they neither chase us nor devour us, but instead, they remain quietly in the background. These visions of transcendence are no doubt impelled by the nagging fear that there is no escape from environmental devastation." People often project a sinister quality onto nature out of fear or from feelings of uncertainties. Perhaps it is also from curiosity about the workings of nature.

The local monster has often been referred to as Hoppie. The influence of Hoppie extended beyond mere fascination, as evidenced by an incident recounted in the August 4, 1894 edition of the *Lake Hopatcong Angler*. According to the publication, not only did individuals claim to have sighted the creature, but one person even attempted to shoot it. The report states, "One man asserts that he hit the head of the supposed serpent with a ball from his .38-caliber rifle, and the bullet rolled off like water off of a duck's back without even making the monster wink." While the article amusingly suggests the presence of a "desperate imagination" in the account, it also jokes about the possibility that the man may have been shooting at a beer keg instead. Nevertheless, the publication of such a story highlights the degree of familiarity and engagement that the local community had with legendary creatures like Hoppie.

A depiction of Hoppie on a coffee mug made by Sandi Astras. *Author's collection.*

Sightings of monsters in Lake Hopatcong have persisted. In 2014, the local headlines were captivated by reports suggesting the presence of a contemporary manifestation of Hoppie. Based on eyewitness descriptions, expert Gerald Andrejca identified a green anaconda as the potential creature in the lake, leading to a swift escalation of the situation and national attention. The incident even found its way into popular culture, with late-night TV talk show host David Letterman humorously linking it to the state's Bridgegate scandal and commenting on the snake's purported ability to consume cows and large governors. Hoppie even had its own Twitter account with four hundred followers.

While the incident may appear humorous in hindsight (and perhaps it was a bit amusing at the time as well), it took on a more serious tone when news publications like the *New York Daily News* reported it under the headline: "Snake on the Loose at New Jersey Lake Is Green Anaconda: Expert," causing a certain level of panic. Subsequently, the police investigation determined that the claim lacked credibility. It's true what they say: all that glitters isn't gold—or in this case, all that slithers isn't a snake. Nevertheless, this episode left a lasting impression.

In his work *Anxieties of Access: Remembering as a Lake*, James L. Smith delves into the multifaceted nature of people's perceptions when gazing into the depths of a lake. Smith provocatively suggests that the act of looking into the lake may evoke not only a visual encounter but also a momentary glimpse of cultural memory and history intertwined with the sedimentary layers of the lake itself. Furthermore, Smith emphasizes that the very act of entering this liminal space defined by the lake entails engaging with the complex array of problems and anxieties associated with this collective memory.

The manifestation of the anaconda in the collective imagination can be understood as a product of memory taking a recognizable form in contemporary society. While a mythical sea creature with a dog-like head may not capture the attention of modern headlines, the representation of a snake as a formidable monster resonates more readily with present cultural sensibilities. The anaconda, therefore, serves as a symbolic conduit through which human feelings and fascination with the unknown aspects of nature can be expressed. It allows individuals to grapple with these fears in a more tangible and comprehensible manner. While the immediate and tangible threat posed by the proliferation of algae in the lake presents a pressing and ongoing danger that may compromise its recreational use, it is perhaps easier for individuals to envision and confront a threat that can be hunted or confronted, reminiscent of the heroic exploits depicted in adventure novels by authors like Rex Beach.

Predicting the Future: Reflections on History, Global Warming and Human Lifestyles in Lake Hopatcong

The late Harvard law professor Christopher Stone introduced a captivating concept of bestowing legal rights on natural entities like forests, rivers and other elements. This pioneering approach offered a novel means of addressing the environmental degradation that is occurring worldwide. While Stone passed away in 2011, his idea lives on. For instance, Lake Mary Jane, which is much like Lake Hopatcong with its irregular shape, filed a lawsuit in 2022.

Located in Central Florida, Lake Mary Jane is facing the pressures of rapid urbanization and growth in Orange County. A proposed development near the lake intends to transform extensive wetlands and natural areas into residential and commercial spaces. In response, Lake Mary Jane, together with Lake Hart, the Crosby Island Marsh and two adjacent streams, has filed a lawsuit in the Florida state court, contending that the proposed development would have detrimental consequences for the lake's well-being. While the judge dismissed it, this case stood as the first instance of a nonliving natural entity seeking remedy on behalf of itself. While the lawsuit sparks debates and conflicting perspectives, it signifies one potential solution to upcoming environmental challenges.

Predicting the future comes with its fair share of challenges. This is like trying to hit a moving target—even if you're closely watching the ball, you never know exactly where it's going to end up. The fantasy writer Laini Taylor once wrote: "There was only present, and it was infinite. The past and the future were just blinders we wore so that infinity wouldn't drive us mad." In the case of lakes and their response to new global challenges, a view of the current state provides valuable insights for anticipating future changes, although uncertainties will inevitably persist.

Thus, history becomes a valuable tool for the study of the environment, offering insights and lessons that contribute to our understanding of the world, individuals and perceptions.

In his 1984 work *Sussex County: A History*, Warren Cummings contemplates the future of Sussex County, recognizing the challenges brought about by rapid development, particularly in relation to water, waste and transportation. Cummings further wonders:

> *But basically, we're right back where we were two-hundred years ago. People poured into Sussex County in the 1780s because it looked to them like a good place to live. They are coming in the 1980s because it still is.*

The trend that emerged in the 1980s has persisted and is likely to continue, making it an aspect that must be acknowledged and effectively addressed.

It becomes crucial to grasp the intricate interplay between climate change and the unique characteristics of Lake Hopatcong to effectively anticipate and manage future environmental changes. The complexity of natural systems, coupled with the inherent unpredictability of future global changes, means that complete certainty is elusive. Therefore, a cautious and adaptive approach is necessary to navigate the uncertainties and challenges that lie ahead.

The future presents qualms not only in terms of environmental changes but also in the evolution of human lifestyles. Throughout history, various modes of transportation, such as the railroad, trolleys and automobiles, have shaped the patterns of human settlement around Lake Hopatcong. The challenge now lies in finding the optimal integration of multiple modes of transportation, considering the efficiency and sustainability of these systems. Whatever this might look like, it is imperative to invest in infrastructure.

Exit 28 to Lake Hopatcong. *Author's collection.*

The construction of Route 80 marked a significant development in transportation around Lake Hopatcong. The process commenced with the introduction of the initial stretch, referred to as the Bergen-Passaic Expressway, in 1954. This segment established a direct connection between the George Washington Bridge in Fort Lee and Route 46 in Ridgefield Park, facilitating more seamless travel within the area. In the 1960s, the section of Route 80 from Ridgefield Park to the Delaware Water Gap was finished. This expansion greatly improved transportation between Eastern New Jersey and the beautiful Delaware Water Gap area. In the early 1970s, the final portion of Route 80, stretching from the Delaware Water Gap to the Pennsylvania state line, was completed. This marked the full completion of Route 80 in New Jersey, enabling efficient travel for the Lake Hopatcong region between the eastern and western parts of the state.

Today, Route 80 carries a heavy volume of traffic, with approximately 160,000 vehicles traveling on it daily. This often leads to congestion, particularly around exit 28, which leads to Lake Hopatcong, especially during the summer months when more people visit. It may be time to reevaluate

The Lake Hopatcong Train Station. *Author's collection.*

the potential of train transportation in efficiently and conveniently moving people in and out of the area. By exploring the possibility of expanding the area's one rail line, the Lake Hopatcong region could benefit from enhanced connectivity and improved transportation options. But in the meantime, let's hope everyone has a good traffic app installed on their phone.

Currently, Lake Hopatcong trains mostly operate exclusively during peak hours on weekdays and during some holiday weekends. This means that those who wish to visit during a regular weekend must rely on their cars. Expanding the rail service would present a practical alternative to driving, easing the traffic congestion on Route 80 and reducing gridlock around the exits. It would provide residents and visitors with a convenient mode of transportation, allowing them to bypass the traffic bottlenecks that often plague the area. An extended railway will also reduce carbon emissions.

To keep the lake healthy, we'll need to keep coming up with new strategies to tackle any unexpected problems that come our way. While daunting challenges are on the horizon, there's no denying the formidable magnetism of Lake Hopatcong will continue to make it an object of desire. This

Peaceful Lake Hopatcong at sunset. *Author's collection.*

stunning body of water has and will continue to be a vibrant symbol of natural beauty and peacefulness. There's nothing like waking up to the sun rising over the lake, casting a golden hue on the calm water, or seeing a blaze of colors mirrored on the lake's surface in the evening. The quiet rustle of trees, the soft lapping of waves against the shore and the chorus of wildlife echoing across the water provide a soothing soundtrack to this picturesque landscape and a calling for people to come and enjoy it.

Lake Hopatcong symbolizes a shared heritage and collective responsibility. Therefore, the future of Lake Hopatcong hinges on joint efforts to nurture and protect this treasure in the face of an ever-changing world. I believe in our ability not only to preserve but also to enhance the splendor and ecological health of Lake Hopatcong. This responsibility has been handed down through generations, from those who encountered or inhabited the area around Lake Hopatcong, dating back to prehistoric times. As we navigate the uncertainties of the future, our current actions will leave a lasting impact on the tomorrows of Lake Hopatcong.

SELECTED BIBLIOGRAPHY

Books and Print Materials

Adamson, Joni. *American Indian Literature, Environmental Justice, and Ecocriticism: The Middle Place.* Tucson: University of Arizona Press, 2003.

Armbruster, Karla. *Beyond Nature Writing: Expanding the Boundaries of Ecocriticism.* Charlottesville: University Press of Virginia, 2001.

Bain, Ann. *Life on a Sussex Farm.* Newton, NJ: N.p., 2008.

Belton, Thomas J. *Protecting New Jersey's Environment: From Cancer Alley to the New Garden State.* New Brunswick, NJ: Rutgers University Press, 2011.

Black, Jonathan. *Sussex County, New Jersey: [Including Its History, the Sterling Hill Mining Museum, the Plaster Mill, the Allamuchy Mountain State Park, and More].* N.p.: Earth Eyes Travel Guides, 2012.

Buell, Lawrence. *The Future of Environmental Criticism: Environmental Crisis and Literary Imagination.* Malden, MA: Blackwell Publications, 2005.

Christensen, Laird, and Hal Crimmel. *Teaching About Place: Learning from the Land.* Reno: University of Nevada Press, 2008.

Coleman, Loren, and Bruce G. Hallenbeck. *Monsters of New Jersey: Mysterious Creatures in the Garden State.* Mechanicsburg, PA: Stackpole Books, 2010.

Cross, Dorothy. *New Jersey's Indians.* Trenton: New Jersey State Museum, 1970.

Cummings, Warren D., Bill Rutherford and Bonnie Rutherford. *Sussex County: A History*. Newton, NJ: Rotary Club of Newton, 1984.

Delaware, Lackawanna and Western Railroad Company and William Henry Johnson. *Summer Excursion Routes and Rates.* New York: Livingston Middleditch Co., 1895.

Ferry, David. *Of No Country I Know: New and Selected Poems and Translations.* Chicago: University of Chicago Press, 1999.

Gingerich, Joseph A.M. *In the Eastern Fluted Point Tradition.* Salt Lake City: University of Utah Press, 2013.

Goller, Robert. R. *The Morris Canal: Across New Jersey by Water and Rail.* Charleston, SC: Arcadia, 1999.

Grumet, Robert Steven, and Frank W. Porter. *The Lenapes.* New York: Chelsea House, 1989.

Harrington, Mark R. *The Indians of New Jersey. Dickon Among the Lenapes.* New Brunswick, NJ: Rutgers University Press, 1966.

Kraft, Robert. *The Lenape-Delaware Indian Heritage.* N.p.: Lenape Books, 2021.

Lake Hopatcong Yacht Club. *Commodore: Hopatcong Historama.* Foreword by Charles R. Rosevear Jr. Newark, NJ: Style Print Co., 1955.

Lurie, Veit, R.F. Birkner, M.J. Gillette, H. Piehler, G.K. Greenberg, B. Greene, L. Hodges, G.R.G. Fea and J.P. Israel. *New Jersey: A History of the Garden State.* New Brunswick, NJ: Rutgers University Press, 2012. https://doi.org/10.36019/9780813554105.

Macasek, Joseph J. *Guide to the Morris Canal in Morris County: A Layman's Working Guide to the Elusive Remains of One of New Jersey's Fascinating Historic Canals.* Morristown, NJ: Morris County Heritage Commission, 1997.

Maher, Neil M. *New Jersey's Environments: Past Present and Future.* New Brunswick, NJ: Rutgers University Press, 2006.

Merchant, Carolyn. *The Death of Nature: Women Ecology and the Scientific Revolution.* New York: HarperCollins, 1989.

Morris Canal and Banking Company. *Rules and Regulations of the Morris Canal.* New York: Printed by J. Van Norden, 1835. www.loc.gov/item/06008995.

Penn, William, Albert C. Myers and John E. Pomfret. *William Penn's Own Account of Lenni Lenape or Delaware Indians.* Moorestown, NJ: Middle Atlantic Press, 1970.

Pomfret, John E. *Colonial New Jersey: A History.* New York: Charles Scribner's Sons, 1973.

Roberts, Russell. *Rediscover the Hidden New Jersey.* 2nd ed. New Brunswick, NJ: Rutgers University Press, 2015.

Seas, Kristen. "Writing Ecologies, Rhetorical Epidemics." In *Ecology, Writing Theory, and New Media: Writing Ecology.* New York: Routledge, 2012.

Smith, Frank E. *Land between the Lakes.* Lexington, Kentucky: University Press of Kentucky, 2014.

Stager, Curt. *Still Waters: The Secret World of Lakes.* New York: W.W. Norton and Company, 2018.

Sullivan, John Langdon. *Refutation of Mr. Colden's "Answer" to Mr. Sullivan's Report to the Society for Establishing Useful Manufactories in New-Jersey upon the Intended Encroachments of the Morris Canal Company in Diverting from Their Natural Course the Waters of the Passaic.* N.p.: 1828.

Thompson, Mary W. *A Summer's Adventure on the Morris Canal (Early 1900's).* Roxbury, NJ: Roxbury Township Historical Society, 1974. Print.

Veit, Richard F. *The Old Canals of New Jersey: A Historical Geography.* Little Falls: New Jersey Geographical Press, 1963.

Vincent, Warwick F. *Lakes: A Very Short Introduction.* Oxford, UK: Oxford University Press, 2018.

Wacker, Peter O. *Land and People: A Cultural Geography of Preindustrial New Jersey: Origins and Settlement Patterns.* New Brunswick, NJ: Rutgers University Press, 1975.

Wichansky, Paul S., et al. "Evaluating the Effects of Historical Land Cover Change on Summertime Weather and Climate in New Jersey." In *New Jersey's Environments: Past, Present, and Future.* Edited by Neil M. Maher. New Brunswick, NJ: Rutgers University Press, 2006.

Databases

Daily Record
Digifind-It.com
Lake Hopatcong Angler
Library of Congress
New York Times Archives
NJ.com
Patch.com

Museum Archives

A Boat, Possibly Dreadnaught, Heading towards Maxim's Boathouse. Hudson Maxim photographs (accession 1996.312). Undated, box 1/folder 7. Hagley Museum and Library, Wilmington, DE.

A Canal Boat, New Jersey. Albert T. photographs (accession 1986.268). Undated, box 23/folder 355. Hagley Museum and Library, Wilmington, DE.

A Canal Lock and Basin in Boonton, New Jersey. Albert T. photographs (accession 1986.268). Undated, box 23/folder 355. Hagley Museum and Library, Wilmington, DE.

Colden, Cadwallader D., Ephraim Beach, DeWitt Clinton and William Davis. *A Report to the Directors of the Morris Canal and Banking Company.* New York: William Davis Jr., printer, 1827. Pamphlets Collection, Published Collections Department, Hagley Museum and Library, Wilmington, DE.

"Family, 1875–1925." Box 1/folder 13 (accession 2147). Hagley Museum and Library, Wilmington, DE.

Hudson Maxim Enjoying Recreational Time Outside of the Laboratory. Hudson Maxim photographs (accession 1996.312). Undated, box 1/folder 7. Hagley Museum and Library, Wilmington, DE.

"Hudson Maxim Papers, 1851–1925." Box 1/folder 18 (accession 2147). Hagley Museum and Library, Wilmington, DE.

"Hudson Maxim Papers, 1851–1925." Box 1/folder 19 (accession 2147). Hagley Museum and Library, Wilmington, DE.

"Hudson Maxim Papers, 1851–1925." Box 1/folder 20 (accession 2147). Hagley Museum and Library, Wilmington, DE.

Hudson Maxim's Car. Hudson Maxim photographs (accession 1996.312). Undated, box 2/folder 7. Hagley Museum and Library, Wilmington, DE.

Hudson Maxim's House and Property on Lake Hopatcong. Hudson Maxim photographs (accession 1996.312). Undated, box 1/folder 7. Hagley Museum and Library, Wilmington, DE.

Hudson Maxim's Stone Boat House with Archway on the Left. Hudson Maxim photographs (accession 1996.312). Undated, Box 1/folder 7. Hagley Museum and Library, Wilmington, DE 19807.

Missiles in Maxim's Home. Hudson Maxim photographs (accession 1996.312). Box 3/folder 4. Hagley Museum and Library, Wilmington, DE.

Newspaper and Journal Articles

Beautiful Lake Hopatcong: What It Is, Where It Is and How to Get There. Lake Hopatcong, NJ: Issued by Lake Hopatcong Chamber of Commerce, 1919. Pamphlets Collection (Pam 97.259), Published Collections Department, Hagley Museum and Library, Wilmington, DE 19807.

Brown, Jeff L. "Shipping Uphill: The Morris Canal." *Civil Engineering* 81, no. 12 (2011): 46–48.

Capuzzo, Jill P. "Hopatcong, N.J.: 'We Call It Lake Life.'" *New York Times*, August 4, 2021.

Carney, Leo H. "The Environment." *New York Times*, July 22, 1984.

Chicago Daily Tribune. "Danger Threatened by Another Dam: Several Towns in New Jersey Menaced by the Pent Up Waters of Lake Hopatcong." June 23, 1889.

Fallon, Scott. "Lake Hopatcong's Toxic Algae Bloom Renews Fight over Stormwater Law Derided as 'Rain Tax.'" NorthJersey.com. July 9, 2019. https://www.northjersey.com/story/news/environment/2019/07/08/lake-hopatcong-algae-bloom-renews-debate-stormwater-law-derided-rain-tax-phil-murphy/1639022001/.

Goldberg, Dan. "Lake Hopatcong's Disappearing Boathouses Take Regional History with Them." *NJ.com*. September 19, 2009. https://www.nj.com/news/2009/09/lake_hopatcong_boathouse_histo.html.

Izzo, Michael. "Lake Hopatcong's Original Sea Creature." *Daily Record*, July 19, 2014. https://www.dailyrecord.com/story/news/local/2014/07/19/lake-hopatcongs-original-sea-creature/12846533/.

Jennings, Rob. "Letterman Riffs on Lake Hopatcong Snake Tale." *New Jersey Herald*, July 26, 2014. https://amp.njherald.com/amp/4030398007.

Kemp, Joe. "Snake on the Loose at New Jersey Lake Is Green Anaconda: Expert." *New York Daily News*, January 9, 2019. https://www.nydailynews.com/news/national/snake-loose-new-jersey-lake-green-anaconda-expert-article-1.1874763.

Kolbert, Elizabeth. "A Lake in Florida Suing to Protect Itself." *New Yorker*, April 11, 2022. https://www.newyorker.com/magazine/2022/04/18/a-lake-in-florida-suing-to-protect-itself.

Lake Hopatcong Breeze. "Change in Directorate of Lake Hopatcong Corporation." August 8, 1922. www.digifind-it.com.

———. "The Indian." July 7, 1906. www.digifind-it.com.

———. "The Interesting History of Lake Hopatcong." August 12, 1922. www.digifind-it.com.

———. "Mt. Arlington Opens Library." August 12, 1911. www.digifind-it.com.

———. "Niles Rights Meeting." October 28, 1922. www.digifind-it.com.

Lake Hopatcong News. "Ye Olde Lake." June 15, 2009.

Mufson, Steven, et al. "Extreme Climate Change in the United States: Here Are America's Fastest-Warming Places." *Washington Post*, August 13, 2019.

New York Daily Times. "New-Jersey." May 27, 1854. www.digifind-it.com.

New York Times. "Hopatcong, N.J.: 'We Call It Lake Life.'" August 4, 2021.

Smith, James L. "Anxieties of Access: Remembering as a Lake." *Environmental Humanities* 13, no. 1 (May 2021): 245–63. https://doi.org/10.1215/22011919-8867296.

Time. "Science: Death of Maxim." May 16, 1927.

Young, Davis A. "Precambrian Rocks of the Lake Hopatcong Area, New Jersey." *GSA Bulletin* 82, no. 1 (1971): 143–58. https://doi.org/10.1130/0016 7606(1971)82[143:PROTLH]2.0.CO;2.

Websites

Chinn, Hannah. "Waging a $13.5 Million Battle against Pond Scum in N.J.'s Lakes." *WHYY*, June 23, 2020. https://whyy.org/articles/waging-a-13-5-million-battle-against-pond-scum-in-n-j-s-lakes/.

CNN. "The World's Largest Lakes Are Shrinking Dramatically, And Scientists Say They Have Figured Out Why." May 18, 2023. https://www.cnn.com/2023/05/18/world/disappearing-lakes-reservoirs-water-climate-intl/index.html.

History of Lake Hopatcong. "Historical Perspective of Lake Hopatcong." http://lakehopatcong.org/history%20of%20Lake%20Hopatcong.htm.

Hopatcong Borough. "Historical Information." https://hopatcong.org/about_us/history/index.php.

Jefferson Project at Lake George. "The Role of Fresh Water Lakes and Reservoirs in the Global Carbon Cycle." https://science.rpi.edu/biology/news/role-fresh-water-lakes-and-reservoirs-global-carbon-cycle.

Koppenhaver, Bob. "Lake Hopatcong." New Jersey Skylands. https://njskylands.com/history-lake-hopatcong.

Lake Hopatcong Historical Museum. https://lakehopatconghistory.com/.

Lake Hopatcong Yacht Club. "Our History—Lake Hopatcong Yacht Club." https://www.lhyc.com/history.

Mammoth Extinctions. "Extinctions." *SFU Museum.* https://www.sfu.ca/archaeology-old/museum/mammoths/extinct.htm.

Musconetcong Watershed Association. "About the Musconetcong." https://www.musconetcong.org/about-the-musconetcong.

Presinzano, Jessica. "Stranger Jersey: The Monster of Lake Hopatcong." *North Jersey Media Group*, October 29, 2018. https://www.northjersey.com/story/entertainment/2018/10/29/stranger-jersey-sea-serpent-lake-hopatcong-nj/1805341002/.

Princeton Hydro. "Mitigating Harmful Algal Blooms at Lake Hopatcong: Largest Application of Phoslock in Northeast." May 11, 2021.

Rensselaer Polytechnic. "Calculating the Role of Lakes in Global Warming." https://science.rpi.edu/biology/news/calculating-role-lakes-global-warming.

Roxburynewjersey.com. "Hercules Factory Explosion, Kenvil, New Jersey 1940." http://www.roxburynewjersey.com/hercules.htm.